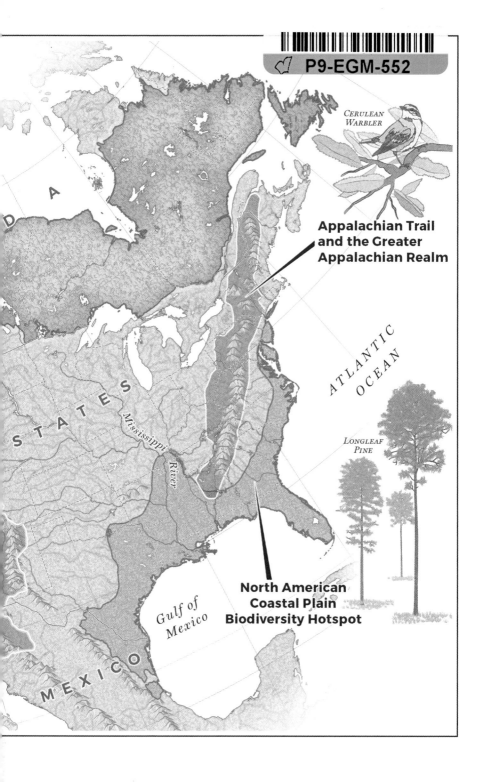

CERULEAN
WARBLER

Appalachian Trail
and the Greater
Appalachian Realm

ATLANTIC
OCEAN

LONGLEAF
PINE

Mississippi

River

North American
Coastal Plain
Biodiversity Hotspot

Gulf of
Mexico

D A

S T A T E S

M E X I C O

RESCUING
the PLANET

—

PROTECTING HALF THE LAND

TO HEAL THE EARTH

Tony Hiss

Introduction by E. O. Wilson

ALFRED A. KNOPF · NEW YORK · 2021

THIS IS A BORZOI BOOK
PUBLISHED BY ALFRED A. KNOPF

Published in the United States by Alfred A. Knopf, a division
of Penguin Random House LLC, New York, and distributed
in Canada by Penguin Random House Canada Limited, Toronto.

www.aaknopf.com

Knopf, Borzoi Books, and the colophon
are registered trademarks of Penguin Random House LLC.

Portions of this work originally appeared in
"Can the World Really Set Aside Half of the Planet for Wildlife?"
in *Smithsonian Magazine* (September 2014).

Library of Congress Cataloging-in-Publication Data
Names: Hiss, Tony, author.
Title: Rescuing the planet : protecting half the land to heal the earth / Tony Hiss ;
introduction by E. O. Wilson.
Description: First edition. | New York : Alfred A. Knopf, 2021.
Identifiers: LCCN 2020026103 (print) | LCCN 2020026104 (ebook) |
ISBN 9780525654810 (hardcover) | ISBN 9780525654827 (ebook)
Subjects: LCSH: Biodiversity conservation. | Nature conservation. |
Conservation of natural resources.
Classification: LCC QH75 .H57 2021 (print) | LCC QH75 (ebook) |
DDC 333.75/16—dc23
LC record available at https://lccn.loc.gov/2020026103
LC ebook record available at https://lccn.loc.gov/2020026104

Jacket photograph by Bradley Wakoff / EyeEm / Alamy
Jacket design by Chip Kidd
Text design by Maggie Hinders

Maps on pages 32–33, 115, 130, and 177 by David Lindroth

Manufactured in Canada
First Edition

For Ann Close, editor and friend

When the blackbird flew out of sight,
It marked the edge
Of one of many circles.

—WALLACE STEVENS

Contents

Introduction

E. O. Wilson

In a dazzling blend of science, travelogue, and history, written in erudite yet pleasurably conversational prose, Tony Hiss has given us as clear a picture of humanity's impact on earth's natural environment as any ever written. Just after I first read the manuscript of this book, I read a special issue of *Time* magazine devoted to climate change whose cover story imagined a world in 2050 that had pulled together at last to deal with the environmental devastation we continue to wreak.

And after reading and reflecting on both, I thought, "Maybe we can save the world after all." Tony Hiss's excellent book helps us understand how it can happen.

I'll take this rare opportunity to summarize the elements of conservation science that will be necessary to achieve the present book's ambitious goal.

Climate change is only the first of at least three crises wrought by humanity destined to inflict major damage on the planet. Also well advanced are the growing worldwide shortage of fresh water and the mass pauperization and extinction of species that lead to the collapse of ecosystems.

A second trio exists that characterizes the biology of the three crises: the ecosystems, such as ponds, bays, forest patches, and, on a small scale, tree holes, root systems, and shelf fungi; next are the species that compose each ecosystem in turn; and finally there are the genes that prescribe the traits of the species that compose the ecosystems.

Species are the key to the future science of conservation practice, as Hiss's report makes abundantly clear. The protection of nature is fundamentally the protection of the species composing the ecosystems. Ecosystems in turn are built from relationships that may have required millions of years of evolution to mature. Ecosystems science is still in an early stage of development. So is its material foundation. Just under 2 million species of plants and animals have been identified and given a Latinized formal name, such as *Homo sapiens*. Yet the total number has been estimated by statistical inference to be, roughly, 10 million. As you read these pages you will learn more, but it is true that in order to create a sound, permanent global world of nature, we need to initiate a "Linnaean renaissance," opening anew the critical scientific initiative begun in 1735 by Carolus Linnaeus to find and name every species on earth.

As an initiative of this magnitude unfolds, other frontiers will be opened. One, as Hiss mentions in this wide-ranging book, will be a far more detailed and sophisticated analysis of soils in different ecosystems around the world, combining chemistry, biodiversity, and every other relevant property of the living and nonliving environment. A key dimension of such a vigorous, comparative soil science will be a full Linnaean account of all microspecies, including protists, bacteria, and viruses.

Tony Hiss's beautiful book shows us ways the planet can thrive. Let the species live!

RESCUING THE PLANET

Getting to Half

On a cold, buckety boat ride early one morning near the top of North America, through a forest that looked endless and only kept on getting bigger, I got to thinking that maybe Henry David Thoreau was only half right when it came to his famous and still-ringing cry that "in Wildness is the preservation of the World." Yes, wildness is the answer, but *people* are missing. Given the state of the planet, if Thoreau were around today, I imagine he'd go the next step and say that "in the People of the Wild is the preservation of the World." Because it is going to take a lot of people to preserve what's still wild, to restore what was once wilder, and to remedy a great calamity in the world.

As a direct result of humanity's destructive actions on the landscape, 1 million species of plants and animals are likely to go extinct, many within the next few decades. A 2019 global assessment, 1,500 pages long, compiled over three years by 145 scientists from fifty countries, makes clear that this mining and undermining of the natural world imperils a millionth-and-first species as well: us. We are "eroding the very foundations of our economies, livelihoods, food security, health and quality of life worldwide," the lead scientist said, leaving only a "very limited" time to turn things around. This deepening emergency, the biodiversity or biocide crisis, was—in pre-COVID-19 days—sometimes referred to as the *other* environmental crisis, to distinguish it from global warming, the damage done to the air, the water, and the climate.

This book is about a way to stave off the mass extinction crisis by keeping life alive.

The idea is to protect far more natural land than any country or continent has in history. According to the World Bank, North America is less than 15 percent protected; the goal is 50 percent over the next thirty years (50 by '50). This may sound unimaginable, preposterous, impossible—in a word, outlandish (literally). It took 150 years after setting up Yellowstone, the world's first national park, to protect almost 15 percent of the continent, and the challenge now is to work at a pace at least twelve times faster.

Though this plan of action is wildly ambitious, I realized on that boat in Canada's Northwest Territories, and elsewhere on other travels in North America, that it is not only doable but is already being done as individuals and groups take it up quietly, bit by bit, in larger and smaller pieces. These people are building on a century-long history of 50 by '50 thinking and are adopting many different approaches—lots of heroes here, many of whom you will meet in these pages, and an upbeat spirit I've had the pleasure to encounter again and again. Here in North America there's still room to spare, and 50 by '50 can move ahead without crowding or displacing or confining anybody because human activities (cities, suburbs, farms, mines, and all the rest) so far account for less than 40 percent of the continent.

At heart it comes down to how to share the earth with other species, and how much of the landscape *not* to change.

According to a calculation by E. O. Wilson, America's foremost biologist, protecting 15 percent of the land guarantees the survival of only a quarter of the species with which we now share the earth. But push that figure up to 50 percent of land and sea, and up to nine-tenths of species will survive. How many species are there? Estimates vary tremendously, from 8.7 million species big enough for us to see, and up to 2 billion or even 1 trillion when you include microspecies, the tiny things we can't see, bacteria and fungi—and far more still if viruses are also counted.

Warning: 50 by '50 is by necessity a bittersweet project. It's a situation where "happily ever after" has to have an asterisk after it. Because of habitat devastation and global warming, some species will be lost no matter what. But the sooner progress can be made, the less regret there will be to carry forward.

There's another layer to 50 by '50, something happening inside people, not just around them. Only about 570 people have had a chance to leave the earth and look back at it, either from the moon or the International Space Station or an orbiting spaceship. Some of them, like NASA astronaut Ron Garan, who spent six months on the ISS, have talked about the experience as life-changing. In "Overview," a video posted on Vimeo that's had more than 8 million views, Garan says, "When we look down at the earth from space, we see this amazing, indescribably beautiful planet. It looks like a living, breathing organism. But it also at the same time looks extremely fragile."

Another NASA astronaut in the video, Jeff Hoffman, who's logged 1,211 hours in space and is now an MIT professor, says, "You go outside on a clear day and it's the big blue sky, and it's like it goes on forever," and yet from space, "it's this thin line that's just barely hugging the surface of the planet." Garan calls it "really sobering" to see how paper-thin this living layer of the planet—the biosphere—is, realizing it's the only thing "that protects every living thing on earth from death, basically." In 1969 Michael Collins was on *Apollo 11,* the first spacecraft to land on the moon. Fifty years later Collins spoke to *The New York Times* about his view of the earth: "I had a feeling it's tiny, it's shiny, it's beautiful, it's home, and it's fragile." Frank White, an American writer on space exploration, calls this new understanding "the Overview Effect." This book is also about opening up to the Overview Effect while still down below.

It isn't so much that we have an affinity for the rest of life, though we do. It's that when danger threatens, it threatens all alike. Here, in the biosphere, a place that has sheltered life for 3.5 billion years, every species shares the same precarious circumstances. More than seven hundred years ago, the Persian poet Jalaluddin Rumi wrote, "Because of necessity, man acquires organs. So, necessitous one, increase your need." In the twenty-first century, need will keep us on our toes at every moment.

In many ways, the mass extinction crisis and the climate crisis are inseparable in the Boreal Forest, since its trees are at once bird rich and carbon rich. Billions of songbirds and shorebirds, some from as far away as South America, come there every spring to nest. The forest and its soil store ("sequestration" is the technical word) billions of tons of carbon that if released into the air would accelerate the warming of the earth. Which is why the Boreal has two nicknames: "North America's bird nursery" and "the Fort Knox of carbon." The North American Boreal Forest, mostly in Canada, partly in Alaska, is the largest and most intact wildness left in the world. On my boat ride in the Boreal I was merely bumping through a gigantic sliver of its breath-catching immensity.

How big is it exactly? Details don't really do it justice when you're in the middle of it way up north, where everything seems bigger anyway, more ancient and undiminished, a place where you're always face-to-face with overarching ecological, hydrological, biological, and atmospheric patterns. But here are a few details: at about 3,700 miles long, the Boreal is a thousand miles longer than the distance from New York to Los Angeles, and north to south (top to bottom) it's up to a thousand miles thick, and altogether it takes up almost as much space as three-quarters of the Lower 48 states. Most essentially, this is a *peopled* wilderness, and has been for thousands of years, its health and abundance maintained, managed, and participated in by more than six hundred Indigenous communities.

I was with some of these people on the boat. Two of them happened to be not just Steves but Steve Ks. The first was a frequent visitor to the Boreal—Steven Kallick, an environmental and human rights lawyer from outside Chicago, who now lives in Seattle and is the director of the International Boreal Conservation Campaign. For twenty years he's been helping build an unexpected—because, really, who would've guessed it?—bicontinental approach to land conservation that has Indigenous groups in the Northwest Territories and across Canada cooperating with Aboriginal allies nine thousand miles away in the Australian outback. A century and a half after Yellowstone, people along this still practically invisible British Commonwealth axis are inventing a post-Thoreauvian, "plus-one" kind of national park, wildnesses that permanently embrace all their plants and animals—plus their human inhabitants. Who (it bears repeating) never left.

The idea here is that these Indigenous Protected Areas, or IPAs, can be set up more quickly than conventional national parks, and they'll be staffed by Indigenous Guardians, paid for by the government, rangers-from-within who already have intimate

The Honorable Stephen Kakfwi   *Steven Kallick*

knowledge of land they've grown up caring for, people who can see the forests *and* the trees and how they work together. Yellowstone was created by taking land away from Native Americans, so this is something else that sets IPAs apart. These parks will never be set up by stripping landscapes from the people who were there first and have looked after them since.

The other Steve K was Stephen Kakfwi, a national figure in Canada. He is a Dene (an Indigenous word meaning "the people") from a village just south of the Arctic Circle and a former premier of the Northwest Territories, which is a little like being governor of a U.S. state. He spent sixteen years as a provincial cabinet member, and he's a former president of the Dene Nation, an organization that's been protecting the rights of the Dene people for fifty years. He's also a folksinger and songwriter; a few days earlier he'd performed live on a Canadian Broadcasting Corporation radio show, accompanying his North Country ballads on a harmonica and guitar. Kakfwi wears his long graying hair in a ponytail, and Kallick resembles a ginger Santa going gray.

The two Steve Ks had come together to talk about what they referred to as the "big chance" for safeguarding the Boreal—an unprecedented opportunity to accelerate the launching of IPAs immediately and over the next ten years throughout the last forest left with more undisturbed areas than any of the other great world forests. Siberia, the Amazon, the Boreal—the Big Three, as they are sometimes known—are all about the same size. But in Siberia roughly 50 percent has been lost, and so has more than 20 percent of the Amazon, where the rate of deforestation is spiking. The Boreal is nearly 85 percent intact, and the ecosystems of this little-known faraway place continue to do indispensable work on the planet's behalf.

In 2017, toward the end of his life, Stephen Hawking predicted that the earth would become uninhabitable within the next one hundred years, forcing humanity to find a new home. But at this point there's no other option for anywhere else to live: of the more than four thousand planets we've found orbiting other suns in the Milky Way, only about twenty are suitable for earth's life-forms, and the closest is more than four years away, traveling at the speed of light. To quote climate activists, there is no Planet B. There were 7.8 billion people on Planet A as of 2020, and the population probably won't start shrinking until 2100, when it will peak at about 11 billion, but the bulk of that growth will arrive between now and midcentury. There's got to be room for the newcomers, too.

Some people think we've already protected the land worth protecting, and anyway there's no need to protect more, since human ingenuity will invent a way out of any crisis. Others find the idea pointless and in vain since we're an inherently destructive species. But in the midst of these doubts, protecting land and sea to save species is gaining unprecedented attention. In 2021, the 196 countries that make up the Convention on Biological Diversity will set a global goal of safeguarding 30 percent of the continents and oceans by 2030. Indigenous people can play a central role in managing another 20 percent, according to a 2019 report, "A Global Deal for Nature." One North American conservation group calls itself Nature Needs Half. E. O. Wilson helped set up the Half-Earth Project. ("Half Earth" is a phrase I came up with while talking to Ed, and he used it as the title for one of his books on the subject.)

So often we turn to the language of war when talking about the environment—the fight to save this, the battle to save that. Even the most prominent Half Earth proponents do this, as I found when I met up with Wilson several years ago in Florida. In the Florida Panhandle, Wilson and I sat on the deep porch of a guest cottage, a house that had half gallons of butter pecan ice cream in the freezer, a Wilson favorite. It was one of those Overview Effect moments, as Wilson's mind ranged far afield, fortified by butter pecan, a conversation that inspired my trip to the Boreal. He talked about taking conservation to the next level—something he thought wouldn't happen without a fight, which he relished. "Battles are where the fun is," he told me, "and where the most rapid advances are made."

Wilson has seen his share of battles. The author of more than thirty books, he is known as the father of sociobiology and hailed as the champion of biodiversity. His now widely accepted theory of island biogeography explains why even large national parks lose species if they're cut off from surrounding landscapes that could let new animals wander in. (Biogeography is the study of what lives where.)

Wilson grew up in and around Mobile, Alabama, and though he's been at Harvard for more than sixty years, he still refers to himself as a southern boy who came north to earn a living. He is courtly, twinkly, soft-spoken, has a shock of unruly white hair, and is slightly stooped from bending over to look at small things all his life—he's the world's leading authority on ants. By the time I met with him, Wilson had earned more than 150 awards and honors, including two Pulitzer Prizes. Now his most urgent task has become a quest to refute skeptics who think there isn't enough left of the natural world to save. "It's been in my mind for years," Wilson said, "that people haven't been thinking big enough—even conservationists."

On the porch, looking out on a lush longleaf pine forest, we took an eons-long view, going back through the 544 million years since hard-shelled animals first appeared. During this time there's been a gradual increase in the number of plants and animals on the planet, despite five mass extinction events. The high point of biodiversity probably coincided with the moment modern humans left Africa and spread out across the globe sixty thousand years ago. As people arrived, other species faltered and vanished. Wilson himself says a "biological holocaust," a sixth extinction, is still possible, and also still preventable, depending on the choices we make now and in the next few decades. "Half Earth is the goal," he told me, "but it's how we get there, whether we can come up with a system of wild landscapes we can hang on to."

Not long after I returned from Florida to my apartment in New York City, I set out to see how we could get to a Half Earth world. I focused on North America, my home continent, the part of the planet that changed conservation by inventing national parks in the nineteenth century and that, in the twenty-first, could again lead the way. I sought out people who consider 50 by '50 practical; I wanted to see the problems they face and the prime spaces where it could work, the pieces of the continent ready to be folded in. I drove through the spectacular jagged peaks of the Canadian Rockies to see a highway that makes room for animals. In Mexico I saw desert land that is no longer desert, but green. I took trips to the U.S. Rockies and to the Northern Sierra, a landscape

John Muir thought was "pervaded with divine light," and to the rewooded New England states, and to northeast Alabama's rugged Paint Rock Forest, and to the almost-flat Delmarva Peninsula, and to my own city block.

I walked around with engineers, trail builders, ranchers, conservationists, biologists, foresters, botanists, and ecologists, among others, and found a kaleidoscopic convergence of perspectives. I also tried to peer through the eyes of other species, to see how animals see places and how their perceptions can expand our own in stunning ways. I caught up on the history of thinking big about landscapes. It was only a hundred years ago, decades before space flight, that a biologist working in what was then the Soviet Union pointed out that the biosphere is gossamer thin. And over the past century ecologists, biologists, foresters, and landscape architects kept coming up with what were essentially Half Earth ideas. As Aristotle noted, well begun is half done.

My companions on the boat ride through the Boreal, the Steve Ks, were Half Earthers before the term existed. Twenty years ago they were already working to protect at least half of the Boreal. Back then this seemed like an unwinnable battle. Today it is finding more and more allies, even across oceans, for this place the planet can't do without and where success is in sight.

The Steve Ks generously shared with me their knowledge of the land, the circumstances of their lives and how they became who they are, and their experience and expertise in reimagining the world for the world's sake. Which made them the perfect guides, and the Boreal the perfect place from which to begin.

Solitudes

A s a boy, Steve Kallick stood among dead fish on the shores of Lake Michigan and felt a sense of dread, not only for the fish but for his place in time. Kallick feared he was born well over a century too late to do right by the environment. He read accounts of the epic seventeenth-century voyage of exploration down the Mississippi River in canoes by the French-American Jesuit missionary Jacques Marquette and Louis Jolliet, the French-Canadian philosophy student who would become a fur trader. Kallick read about how wide and beautiful the Mississippi was as it flowed past both forest and prairie, how Marquette and Jolliet were welcomed in each native settlement they encountered. They couldn't go fifteen minutes without seeing an abundance of wildlife, mesmerizing to Marquette's European eyes. They describe hauling in a catfish with the head of a tiger, the nose of a wildcat, and whiskers.

"Where are the Indians?" Kallick remembers asking his mother. "Where did they go?" In the ways history was presented to a young boy growing up in Illinois, it all seemed so achingly

Marquette and Jolliet on the Mississippi, 1673

long ago. The American frontier, the westernmost edge of settlement, closed even before 1900, and the canoe trip of Marquette and Jolliet had been two hundred years before that. "It left me with an enormous sense of loss," Kallick says.

As a teenager, Kallick's life took an abrupt right turn when he looked north instead of west. He'd always thought he'd be a Major League Baseball player, and came close to trying out for the minors, but eventually he began toying with the idea of a high-paying job on the Alaska pipeline. Salaries offered in the 1970s were the equivalent of a quarter of a million dollars today. When he was seventeen he climbed into an old VW bus, and he and his best friend drove to Fairbanks. When he got there, the nonstop boomtown partying overwhelmed him, more scary than enticing—maybe college or even law school wasn't such a bad idea after all. The two young men spent the rest of the summer driving through Alaska and into Canada, just looking around. Then one day, at the end of a mountain road, they realized they were at the very end of the entire North American road system. Everything that stretched off to the horizon to the north and the east was still wild, as it always had been.

"I had my epiphany," Kallick says. "I'd found the never-conquered frontier, and I was on the edge of it. And my thought was, This needs to be defended. It's not too late—I've just been looking in the wrong direction. History doesn't only spread from east to west. All the things we did wrong the first time, to the land and to the people who were already there—here in the North there's a chance to not let this happen. It can be like going back in time. Doing it differently."

Do-over country, a reset, a place where the inevitable is not inexorable, a place Marquette and Jolliet could recognize, one that still conjures up the special glow of their first glimpse. Not that our river voyage was a historical re-creation. The boat we were on wasn't a canoe, but a trim twenty-two-foot aluminum fishing boat with a powerful outboard motor. The boat's owner, Joe Grandjambe, was the captain, and his wife, Angela, who had a cell phone, came along too—they're from Fort Good Hope, Steve Kakfwi's village, and longtime friends of his. Kakfwi had a waterproof orange plastic case that held a satellite phone, or sat phone, which, because it connects to global satellites and not to cell phone towers, lets you call for help from anywhere in the world. Kakfwi mentioned that he was considering buying a wristwatch-sized version that sends out your precise location every two minutes—"As long as you can see the sky," says one review, "it'll work."

The river we were on was the Mackenzie, sometimes called the Cold Amazon. Like the Boreal, this river is a domain where adding "-er" and "-est" to words about size doesn't begin to convey the scale at which things operate, or the sense of uninterruptedness that enfolds you once you leave shore. It's the second-biggest river system in North America, surpassed only by the Mississippi. In places the Mackenzie can be two and a half miles wide, and if you include tributaries, it's more than 2,600 miles long. All told, the Mackenzie basin drains an area about the size of Alaska. It's flanked to the west by a mountain range, also called the Mackenzies, a northern extension of the Rockies, with 9,000-foot peaks, a 3,300-foot deep canyon, and a waterfall twice the height of Niagara.

Part of its uniqueness is that it flows north—making it a river that "turns its back on America," said F. R. Scott, a Canadian poet. In yet another challenge to a two-way, east-west mind-set, there's a continental divide two hundred miles above the U.S. border. This one's a three-way split, technically a hydrological apex, atop Snow Dome, a peak 11,322 feet high near Banff National Park, where melting snow can flow in any of three directions—west toward the Pacific, east to the Atlantic via Hudson Bay, or, like the Mackenzie, up north. Meaning actually down north—that is, downhill from Snow Dome to the Arctic Ocean, which is as far away as New York City is from Omaha, Nebraska.

Floating the wrong way wasn't exactly disorienting, because that word literally means being detached from a sense of where the east is. Whereas on the boat, what was ebbing away was my own just-discovered sense of southness—replaced with what Glenn Gould, the Canadian pianist, memorably called "The Idea of North." This was the name of a CBC Radio documentary Gould worked on half a century ago, exploring concepts of solitude and Canadian identity.

We moved along the Mackenzie at a steady clip, covering a hundred miles in eight hours—it felt like a fast canter as Joe's boat bounced off whitecaps. Overhead and perpendicular to the river were long, tightly packed rows of gray clouds, like tray after tray of gray baguettes that just kept the sun from breaking through. Beneath these clouds, on either bank of the river, came the spiky verticals of the forest itself—dark, bluish-green lines of black spruce, trees that look to be almost all trunk, tall, narrow, pointed, straight up-and-down, and crowding close together like clusters of tightly furled umbrellas, interspersed with clumps of birch and aspen made golden by the cool nights of early fall. To the right of the gray-green river were bare, grayish-tan rocky ridges, some with abruptly sheer cliffs crumbling into rubble at their base. The cliff seemed to advance toward the river, then draw back again at irregular intervals.

To keep to the best channel, we zigzagged between banks. Each time we swept around wide bends, I'd see countless receding panoramas of black spruce daubed with the fast-growing bright birch that springs up in spots leveled by forest fires. We skirted two dancing, boiling rapids, funneled through the Ramparts, a spruce-topped, 130-foot-high limestone canyon seven and a half miles long, and never once did we pass a city, or village, or catch a glimpse of anything like a dam, bridge, fence, smokestack, power line, road, or even a rudimentary trail.

This is an angular landscape, not a rounded one. Until a century ago most Canadians had no idea what it looked like. But in the 1920s a group of young upstart Ontario artists, contemporaries of Edward Hopper and Georgia O'Keeffe—the Group of Seven, they styled themselves—turned their back on what they called the "cow school," the prevalent practice of painting placid pastoral southern Canadian scenes that looked like the English or French or Dutch countryside. They embraced instead the North's unfamiliar colors and skies and the stark, elemental shapes of its hills and mountains—what their leader, Lawren Harris, called the "singing expansiveness and sublimity" of that landscape. In their paintings, everything remote has an immediacy, "as though," the Canadian writer Mitchell Gray put it in *Canadian Geographic*, "the viewer were peering from the flap of a tent in the wilderness."

The Group of Seven wasn't welcomed by all—an old woman told one of them, "It's bad enough to live in this country without having pictures of it in your home." But thirty years later they were credited with helping define Canadian distinctiveness, and Canadians who grew up in the 1950s remember their prints in schoolrooms across the country.

If the luminous, sensitive paintings of the Seven, many of them twentieth-century masterpieces, have a drawback, it's that they show at best a "Half North." The gorgeous, harsh, as-it-ever-was landscape is an empty place, with lakes and trees but no people, and devoid of animals. (Thus making it a "fourth day of creation" place: in Genesis, fish and birds only appeared on day five, beasts and humans on day six.) These starkly beautiful paintings seem to assume that an untouched landscape has to be an uninhabited one.

Far from the twisted reach of crazy sorrow

The very idea of a blank slate is a cornerstone of what is referred to by historians as "settler colonialism." In a controversial 1992 decision, the High Court of Australia abandoned this view, the legal doctrine of *terra nullius* ("no one's land"), which had been used to justify England's sovereignty over and ownership of Australia. In 1788, a British fleet with more than seven hundred convicts aboard landed in New South Wales and claimed this "no one's land," despite the presence of more than three-quarters of a million Aboriginal people representing up to four hundred nations.

The old thinking went that if the people in a place hadn't visibly reshaped it—"had not yet mixed their labor with the earth in any permanent way" (typically by farming it)—then a place remained "undifferentiated in time and space," as Australian historian Alan Frost explains. New South Wales, said a 1787 London newspaper, was a "Virgin Mould, undisturbed since the Creation," echoing the words of Captain James Cook, who, after sailing along the Australian coast, reported that "the Industry of Man has had nothing to do with any part of it." The presumption: a place untouched by Europeans is a place awaiting fulfillment.

A presumption that even Hugh MacLennan, one of Canada's best-known authors, brought to the Mackenzie River Valley. During World War II, according to MacLennan's biographer, Elspeth Cameron, his wife told him that even though the country produced the Group of Seven, no one would ever truly understand Canada until it had a literature of its own. At the time, 98 percent of the books in the country were by British or American authors. MacLennan then wrote *Two Solitudes*, a powerful novel about the bitter and unresolved tensions between the founding European populations, or races, as they called themselves, English-speaking Protestants and French-speaking Catholics. The title came from Rainer Maria Rilke—"Love consists of this, that two solitudes protect and touch and greet each other"—and also it referred, MacLennan said, to the nations then at war. He saw Canada itself as "a bridge with the ends unjoined."

On a postwar assignment for *Maclean's* magazine, MacLennan toured the Mackenzie River Valley—then as now with a population of ten thousand—and in 1961 confidently predicted, "In the year 2061 there will be at least 3 million people living in the Mackenzie Valley. There will be hospitals, schools, and at least two

universities established on sites overlooking this cold, clean river."
A third solitude, the First Nations who called the valley home and
understood it, remained wholly invisible to this gifted, sensitive
writer, who called the countryside "a land emptier than the sea!"
And a fourth solitude also went unnoticed—the undiminished
abundance of the landscape the First Nations have safeguarded.

What's changed since 1961 to keep the Mackenzie Valley unchanged in so many ways? Why aren't 3 million people living
there? For one thing, the aftermath of "Canada's Brown Holocaust," also called "a national crime" and "the horror of our history." In 2008, the Canadian prime minister told Parliament and
representatives of the First Nations, "The government of Canada
sincerely apologizes and asks the forgiveness of the Aboriginal
peoples of this country for failing them so profoundly. We are
sorry." This was followed by the Truth and Reconciliation Commission of Canada to seek restorative justice for a century of cru-
elty that didn't end officially until 1996. Maybe there had been no
slaughter, no wars, no Little Bighorn, no Trail of Tears, and most
First Nations weren't pushed off the land, but the government
had systematically and chillingly set out "to kill the Indian in the
child."

About 150,000 First Nations children, many from several
generations in the same families, were robbed of their childhood,
each one by law confined for ten years, beginning when they
were seven, to "residential schools" run for the state by Christian churches. (Some children were as young as four and impris-
oned for thirteen years.) Once there, the children were neglected,
abused, and often starved; as many as six thousand children
died of tuberculosis or other diseases. The prevailing theory, as
expressed by a Canadian cabinet member, was that "if you leave
them in the family, they may know how to read and write, but
they will remain savages." Bev Sellars, author of a searing best-

selling 2013 memoir, *They Called Me Number One* (she was never called by her Indigenous name), wrote that she didn't learn to tell time until sixth grade, but by then had already been repeatedly beaten with "the strap" (a strip of leather cut from a conveyor belt) and had absorbed the message that "I was inferior to White people."

The principal of her school, later a bishop, eventually went to jail for multiple acts of sexual abuse. Sellars never expected to live beyond her teens. For years she suffered from nightmares and migraines. Finally she found something that gave her the strength to endure—self-help books, and having a baby she adored. Unlike Sellars and her mother and grandmother, the child was never taken to a residential school; by then the system had been abolished. Sellars became a lawyer, a historian, and an elected chief of the Xat's̄ull First Nation.

Truth and Reconciliation—the name came from the post-apartheid commission Nelson Mandela set up in South Africa. In the Northwest Territories the dynamic might better be described as Remorse and Resurgence, as the ones apologizing became more attentive to the people they'd wronged. According to *Historic Trauma and Aboriginal Healing*, a booklet put out in 2000 by Canada's Aboriginal Healing Foundation and cited by Sellars, it took Europeans four decades to heal the social breakdowns that followed the devastating outbreak of the bubonic plague— the Black Death—in the fourteenth century. The hope is it won't take that long in the Northwest Territories.

Among residential school survivors—pretty much any First Nations Canadian born before 1975—there has been no single way to move beyond despair. On our boat ride, Angela Grandjambe told me she endured school by reading a new book every day. Stephen Kakfwi, who got beaten by a nun who told him he was bad and dark and looked just like the devil, read poetry (Tennyson and Wordsworth), quotes from Lincoln, and the lyrics to new songs (he couldn't listen to the songs themselves because there were no radios or record players). A couple of lines from Bob Dylan's "Mr. Tambourine Man" have stayed with him ever since he first saw them: *"Far from the twisted reach of crazy sorrow / Yes, to dance beneath the diamond sky with one hand waving free."* "Giving up one minute, defiant the next," he says. "I thought, 'I could suddenly change into what I wanted to be.'"

If any good at all came out of the residential schools, Kakfwi says, it's that a band of people from the final generation of incarcerated students stayed in touch and became a network of young First Nations chiefs, spokespeople, and leaders. "Though now we're elders," he adds. Prominent members of the network include Sellars, Kakfwi, and the Honorable Ethel Blondin-Andrew, a Mackenzie Valley Dene who ran away three times from her residential school and, in 1988, was elected to the Canadian Parliament as the first Aboriginal woman MP. Over the next seventeen years she joined the Canadian cabinet, traveled with parliamentary delegations to the Mideast, accompanied Lindsey Graham and John McCain when they visited the Northwest Territories, and went to Bolivia with Al Gore. All of it fun, she says, but with an underlying purpose—helping people back home "start exercising our rights."

Something to learn

In 1972, twenty-seven Canadian and American energy companies announced plans for the Mackenzie Valley Pipeline, an $8 billion, 2,400-mile-long natural gas pipeline, said to be the biggest project in the history of free enterprise, to bring natural gas from Alaska down to southern Canada, along a route paralleling the Mackenzie River. It was immediately clear this would set in motion a permanent "southernization" of the valley, the sort of thing Hugh MacLennan had looked forward to in 1961. To build and maintain a pipeline you'd need new roads, airports, and towns. It was clear, too, that this was entirely a "Second Nations" project, so to speak, developed without consulting the third and fourth solitudes: the people of the North and the profusion of living things on the land. The lines were drawn for a Half Earth–type battle.

It was at this point that Kakfwi, then in his early twenties, "burst into national prominence," as *The Canadian Encyclopedia* puts it. Dene elders, he says, "asked me to tell the world who we are, the people of the earth— *'ne'* means the land, the earth. We have no dominion over it, because we're a part of it, but we're the ones paying careful attention to it. Just as Indigenous people and all human beings have rights, so does the earth. Ever since our first contact with Europeans, other people have had their own plans for us, for it. Time to take back control."

Kakfwi toured the country, explaining Dene opposition to the pipeline. It was only the second job he'd ever had, calling it his political badge of honor. His was one of the forceful voices heard by the Berger Inquiry, a royal commission set up in 1974 to assess the impact of the pipeline, led by Justice Thomas Berger from the Supreme Court of British Columbia. In Canada it's considered the most influential inquiry ever convened because of the way it balanced the thinking of the North and the South, giving equal weight to proponents of the pipeline, who likened it to "a thread stretched across a football field," and those for whom it was "a razor slash across the *Mona Lisa.*"

Unlike the canvases of the Group of Seven, the picture of the North the commission drew was filled with the people of the North. Over a three-year period, Berger, accompanied by reporters, CBC Radio correspondents, and a film crew, visited and held hearings in seven languages in thirty-five towns, listening to Dene, Inuit, and Métis, as well as to non-Indigenous northerners, "in log cabins, village halls, besides rivers and in hunting and fishing camps," a colleague, Ian Waddell, later a member of Parliament, remembered.

Berger compiled more than forty thousand pages of transcript. It was the only way, he said, to put First Nations on "something like an equal footing with industry." No one in authority had done anything like this before. As Berger later described the process, "I went to all the Aboriginal communities. I said, 'OK, I'm here. I'll stay as long as you want me to. I want to know what you think about this. You live here, it's going to affect you, it's your future.' And Canadians grew used to the idea that Aboriginal people did have something to say and they were going to say it."

When it was Kakfwi's turn to testify, he made fun of Alexander Mackenzie, the eighteenth-century explorer for whom the river and the valley are named. Mackenzie described the Dene as "meager" with "scabby knees"—so Kakfwi described Mackenzie as "this strange, pale man in his ridiculous clothes" and said the Dene would "never understand why their river is named after such an insignificant ungrateful fellow."

Kakfwi told the commission, "We have been the owners of this land long before the white man came and formed Canada, but because we never wrote our views and values of this land down on paper before, the Government seems to think we never thought that way before. . . . Our reality is that this is our land, that we are a nation of people and that we want to live our own ways. Our reality is that the pipeline is just a poorly masked attempt to overwhelm our land and our people with a way of life that will destroy us. Our reality is that all of the 'help' your nation has sent us has only made us poor, humiliated and confused. Our reality is that we are in great danger of being destroyed. Our reality is that there is a very simple choice—Dene survival with no pipeline, or a pipeline with no Dene survival."

Berger wrote a powerful letter with his report, one that still resonates today:

> We are now at our last frontier. It is a frontier that all of us have read about, but few of us have seen. Profound issues, touching our deepest concerns as a nation, await us there.
>
> The North is a frontier, but it is a homeland too, a homeland of the Dene, Inuit and Metis, as it is also the home of the white people who live there. And it is a heritage, a unique environment that we are called upon to preserve for all Canadians.

Near the beginning of the report he stated:

> What happens in the North, moreover, will be of great importance to the future of our country; it will tell us what kind of a country Canada is; it will tell us what kind of a people we are. In the past, we have thought of the history of our country as a progression from one frontier to the next. Such, in the main, has been the story of white occupation and settlement of North America. But as the retreating frontier has been occupied and settled, the native people living there have become subservient, their lives molded to the patterns of another culture.
>
> We think of ourselves as a northern people. We may at last have begun to realize that we have something to learn from the people who for centuries have lived in the North, the people who never sought to alter their environment, but rather to live in harmony with it. This Inquiry has given

all Canadians an opportunity to listen to the voices on the frontier.

The commission urged scrapping the part of the pipeline planned for northern Yukon that would threaten the integrity of a "wilderness of incredible beauty." Beyond that, it recommended that the entire project be delayed ten years so that First Nations land rights issues could be settled. Otherwise, Justice Berger's report said, "the social consequences of the pipeline will not only be serious—they will be devastating." The Berger Inquiry has come to be known as "Canada's Native Charter of Rights."

Eternal reminder

Before the first of the river's big rapids, Joe Grandjambe slowed, steered for the shore, and landed us on a small rocky beach below a sharply rising cliff, Carcajou Ridge, forested but with long patches of bare rock. We hopped off to stretch our legs. In a small muddy spot was a single sharply etched black bear paw print so fresh it looked only minutes old. The wind, the current in the river lapping at the beach—nothing else stirred. I was thinking how isolated the place seemed when Joe pointed up at the bare rock overhead and told us a Dene story about a spirit, and giant beavers, and a wolverine.

A spirit sent by the Creator to bring order and rules for living in harmony, Yamoria, killed three of the last giant beavers that were then menacing the world and killing Dene hunters. Yamoria left their brown, bloody pelts to dry on a high hill, and a wolverine stole some of the meat. As an eternal reminder that stealing is not a virtue, Yamoria turned the wolverine into rock.

Back on the river, the bare patches of rock on Carcajou Ridge had, yes, without stretching things, a striking resemblance to a crouching or sleeping wolverine. In another place and moment, the story might've seemed quaint, fanciful. But the hill where Yamoria left the beaver pelts, Bear Rock mountain, 1,300 feet high, is a hundred miles away, many hours distant even on a boat with a powerful engine. To this day one bare face of Bear Rock has three large brown-red oval spots one hundred feet tall, visible for miles. From fossils, archaeologists have established that *Castoroides*, a now-extinct giant beaver the size of a bear, up to eight feet long and weighing more than 250 pounds, was active in the area as recently as ten thousand years ago.

A story of seeming simplicity—compressed, portable, memorable, inheritable—that can be held in the mind without effort. The Dene look at the world in a way that encompasses a place that is not just physically vast (the wolverine rock one hundred miles from the beaver-pelt rock) but temporally vast (at least ten thousand years old). The story Joe shared is a kind of shorthand, a way of understanding the entire sweep of a deep-time landscape, vast and untrammeled as it is, through generations of storytelling, and keeping it close to heart. A young Dene leader, a cousin of Steve Kakfwi, said back in the days of the Berger Inquiry, "There is no word in our language for wilderness, for everywhere we go it is our home."

Gaps

For people not used to it, there's something unsettling about a roadless area. In the upper Mackenzie Valley, if you don't have access to a boat like Joe's, the only way to get from one village to another is by plane. There's jet service to one village on Canadian North, an airline owned by the Inuit, and there's local village-to-village service on a couple of snug little turboprop Beechcrafts run by North-Wright Airways, owned in part by several First Nations communities—homey flights where the jokey pilot likes to say things like, "We'll be on our way to Vegas and Palm Springs as soon as I get the rubber band wound back up." In winters, once the lakes and rivers freeze, there are ice roads, but otherwise the old punchline holds sway: "How do I get to—" "Oh, you can't get there from here."

Being in a car on a road that's part of the interconnected highway system is the everyday situation of all "southerners"—people who live below the sixty-third parallel, about eight hundred and forty miles above the U.S. border, near where the roads stop. Maybe it's something you notice only in its absence—I'd never been beyond the reach of the roaded globe before. But if you've grown up with roads, when you're on one, there's a tug of the familiar, a faint caravan effect that tells you you're always in the presence of the known, whether it's places up ahead or people back home. It could be only a back-of-the-mind tendency, but on a road there's a kind of devaluing of where you are at the moment, making it seem emptier, more drained of meaning, a place "unstoried." When roads arrive they bring a mind-set, or at least the whiff of one, an expectation and assumption, however faint, that development won't interrupt anything essential. And not only roads. When arbitrary lines get affixed to a landscape, they can usurp the landscape itself.

In 1785, the newly independent United States was suddenly a lot bigger than the original thirteen colonies and needed to raise money quickly to pay war debts. Land beyond the Ohio River got sold off to create what was called "an internal empire of small farms." It was Thomas Jefferson, then in Congress and not yet president, a champion of the "yeoman farmer" ("small land holders," he said, "are the most precious part of a state"), who suggested a rectilinear cadastral survey—a series of straight lines at right angles—arranging the land into easily marketable pieces as the speedy solution to the problem. It's called the "Jefferson grid."

Traditional English land surveys, the norm on the East Coast, used the metes and bounds system, which required a lot of work. Generally the edges of any particular property were natural or human-made features recognizable to all—a brook, rock, tree, stone wall, barn—which surveyors then connected with as many lines as necessary, recorded on a map as a unique, individual, irregular shape.

What may have prompted Jefferson was slick new 1700s technology: an English clockmaker created a chronometer that allowed for accurate east-west measurement, even at sea where there are no natural features to take bearings from (astrolabes and quadrants for north-south distances had been around for centuries). Jefferson applied these oceangoing longitude and latitude measurements, in effect treating the land like the sea, marking it off mathematically into thirty-six-square-mile townships, which then got subdivided and subdivided again into smaller square lots called quarter sections. All these straight lines ran unyieldingly up hill and down dale, ignoring the shapes of mountains, valleys, forests, prairies, and rivers—and their Native American inhabitants.

There are now nearly 1.5 billion acres of Jefferson grid in thirty states, a rational, efficient, "single, hyper-rigorous design," it's been called. Imposing the grid signaled starting-from-scratch time in America—back to square one, if you like. Except the new squares had no context or continuity. Richard Manning, a grassland historian, wrote that the Midwest prairies got turned into "a blank slate needing only lines, plows and bags of European seeds." Before filling in the blanks, the grid was blanking in the fill.

"I have no use for the man"

Joe stopped the boat midriver to indicate the spot on the shore where a group of Dene rescued Alexander Mackenzie, later Sir Alexander and now considered a National Historic Person—one of the "discoverers," Hugh MacLennan said, "whose names will never die in the history of western exploration." In 1793, more than a decade before Lewis and Clark, Mackenzie led the first crossing of North America—but when the Dene came upon him several years earlier, he was only twenty-five, a fur trader in search of the Northwest Passage, a waterway across the continent that (if it actually existed) would cut the time to sail to China.

"There's so much history here," Joe said, "and the only part I don't like to tell is Mackenzie. It eats away at me. It's so much worse than what Kakfwi told the Berger Inquiry, about Mackenzie being strange and wearing ridiculous clothes. When the Dene found Mackenzie, he was stopped in his tracks. Several in his party were dead. The Dene brought the survivors to caves and kept them alive the whole winter, and I'm not sure why they bothered. Mackenzie was free with his insults of the Dene and never mentioned their help in his diary. The Dene made possible his later exploits—and then the Canadians went and named the river after him. The river wasn't even the passage he wanted it to be, since it bends north and empties into the Arctic Ocean, not the Pacific, and supposedly the name he thereafter gave it was the 'River of Disappointment.' I have no use for the man."

"But what happens when a story is incorrectly told, or when a story is missing altogether?" ask the authors of *Hidden in Plain Sight: Contributions of Aboriginal Peoples to Canadian Identity and Culture*. It appears as an epigraph to Bev Sellars's second book, *Price Paid: The Fight for First Nations Survival*, written to "fill in some of the gaps in the history of Native-newcomer relations in Canada." Canadians, she notes, need to become "explorers of the explorers." *Price Paid* makes it clear that Mackenzie's is not an isolated case, and after the "silent voices" get restored to the story of Canada, the age of exploration looks more like an Indigenous taxi service with health care providers on call.

"North, Central and South America were fully mapped out to take any explorer anywhere they wanted to go," Sellars told a CBC interviewer. There was never any opening up of the West because the West was already open. There was never any hacking through a pathless, forested wilderness, because the newcomers—the men whose names will never die—were always in the care of Aboriginal guides who knew the ancient routes, and shared food and medicines, and who knew how to fix the canoes, kayaks, dogsleds, and snowshoes they were providing, because they had invented all of them.

Opportunity for permanence

Charter a small plane to fly over the Boreal Forest, and you may not see a moose a minute, but chances are you'll see one or several every five or ten minutes—their numbers are strong and growing. This despite a devastating 2020 survey by the World Wildlife Fund, which found that within the last half century, 68 percent of the bigger wild animals on the planet—mammals, birds, fish, reptiles, and amphibians—have disappeared, either killed outright or because their habitats were destroyed for farms, mines, towns, and factories. These animals haven't gone extinct; there are just far, far fewer of them. In southern Canada, according to World Wildlife Fund Canada, there's an 83 percent loss.

Meanwhile, one outstanding characteristic of the Boreal now, and a century ago, and farther back still, is the overflowing abundance of animal life, predators and prey—grizzly bears, black bears, cougars, wolves, and wolverines; wood bison, caribou, elk,

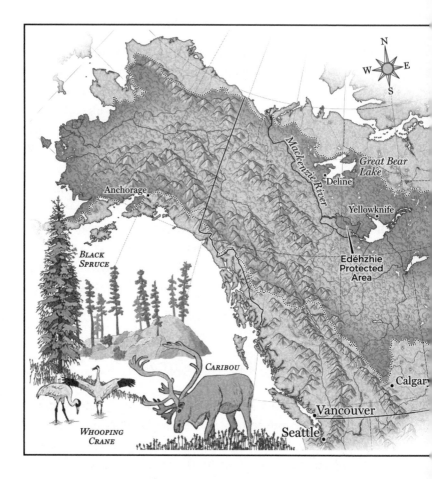

moose, beavers, snowshoe hares. There are 85 different mammals in all, along with around 130 species of fish, including ancient, giant lake trout that can weigh eighty pounds and live sixty years; the water's so cold they gain only about a pound a year.

Also there are more than 300 species of birds, along with something like 32,000 different kinds of insects, including black flies and mosquitoes, that provide the birds with summer-long feasts. The Boreal is "North America's bird nursery" because up to 3 billion birds fly there every spring to raise another generation;

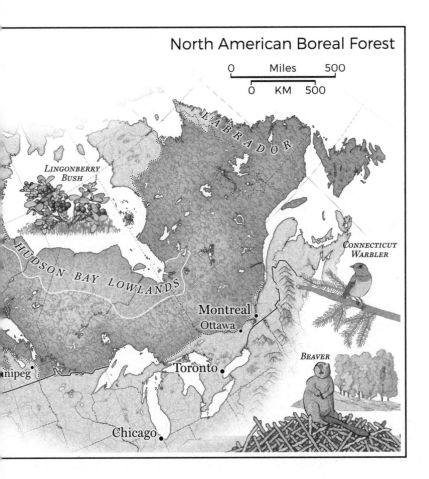

North American Boreal Forest

0   Miles   500

0   KM   500

LABRADOR

LINGONBERRY BUSH

HUDSON BAY LOWLANDS

CONNECTICUT WARBLER

Montreal

Ottawa

BEAVER

nipeg

Toronto

Chicago

every fall, 3 to 5 billion birds, parents and offspring, fly south again. Unchanged abundance—and with it an undiminished vagility, meaning the ability of wide-ranging animals like wolves and caribou to move freely and without restriction across a landscape that isn't now and never was too small for them.

The green forest of the Boreal is also called a "forest of blue," the continent's greatest outpouring of water, with a quarter of the world's wetlands and four of the planet's largest lakes. "Great Lakes" is a name that stops short of describing what's there; they are merely one end of a sinuous arc of more than a million lakes and ponds, culminating in two lakes bigger than Lake Erie and Lake Ontario. This glittering sweep of fresh water, much of it connected by undammed, free-flowing rivers, swings north and west from Lake Superior for two thousand miles until it reaches the Mackenzie Valley and finally the Arctic Ocean. Stronghold, showcase, refuge: some of the original reasons for wanting to protect the Boreal.

"Wild places are facing the same extinction crisis as species"—this was the conclusion of a 2018 *Nature* article by Australian, Canadian, and U.S. conservation biologists. They pointed out that a century ago, only 15 percent of the earth's surface was farmed or used for grazing livestock, and that today people make use of more than 77 percent of the land, except in Antarctica.

The integrity of all ecosystems is at stake. Why? Intact ecosystems, like the Boreal, do more than safeguard wildlife and abundance. They also protect the further evolution of biodiversity, our best opportunity for permanence, since it's a link to billions of years of growth, the crucible within which life can withstand natural change or adjust to it. But evolution and biodiversity can only happen if wilderness areas are still here and can stay large enough, long enough; wilderness size and evolution seem bound together. In their *Nature* article, the Australian and North American biologists put forth what they describe as a "bold yet achievable target"—to conserve "100% of all remaining intact ecosystems."

Ys and CALs

What makes an ambitious goal achievable? Even in its shrunken state, there's still so much wilderness—something like 7.5 billion acres globally, with almost a billion and a half in the Boreal. It might help to de-dazzle it, make it more ordinary, less intimidating. Cutting size down to size—it's been done before.

One way to let our minds take in more territory is to coin shortcut words that encompass the inconceivable: astronomical unit, light-year, googol. An astronomical unit, or AU, is a yardstick for the solar system. It's 93 million miles from the earth to the sun, but only 1 AU; Mars is 1.5 AU from the sun. Beyond the planets, AU gets cumbersome—it's 268,770 AU to the nearest star but just 4.25 light-years (the distance light, which moves at 186,000 miles a second, can travel in a year). "Googol," a deliberately silly word, was invented by the nine-year-old nephew of a mathematician who asked him to name the number that's one followed by one hundred zeros; but nevertheless, mathematicians like to say, it's no closer to infinity than the number one is. (The founders of Google, the search engine, chose the name to show they could point people to huge quantities of information, accidentally misspelling it, keeping the name even after realizing their mistake.)

One night in Fort Good Hope, I brought this up with Steve Kallick, someone accustomed to taking hugeness in stride. A shimmering band of green arced across the sky from horizon to horizon—the aurora borealis (northern lights)—and we came up with a couple of "continental units," or CUs, yardsticks for the Boreal Forest. In our forest math, we would use Ys and CALs. "Y" would stand for Yellowstone, the first large protected area, which has a convenient, round-number size of just over 2 million acres, while "CAL" is short for California, the third-largest state and a place whose size is also easy to remember, about 100 million acres. In this rough CU calculation, 1 CAL = 50 Ys.

North American parks are often referred to, confusingly enough, in terms of "Delawares" and "Rhode Islands," the two smallest states. Yellowstone is almost as big as those two states put together. Within the Boreal there are lakes as much as ten times larger than Rhode Island. National parks in the Lower 48—including Yellowstone—are spread out over about 20 million acres. That's 10 Ys. Alaska's parks are bigger, led by a couple of 4 Ys, but even so, the full complement of the National Park Service's four-hundred-plus units adds up to less than a single CAL. While in the Mackenzie Valley, just the first four Indigenous Protected Areas in the works total a whopping 14 Ys, and in the mountains to the west there are plans for another 18 Ys. To the east of the valley, between it and Hudson Bay, an area called the "wild heart" of the Boreal, there is almost a CAL of intact land and prospects for parks that would protect 32 Ys.

In a city, as it grows, a single skyscraper becomes part of an ever more compelling skyline. Perhaps in the Boreal that will be the outcome for a couple of Northwest Territories parks. One's about a 5 Y: Wood Buffalo, Canada's largest park and also the

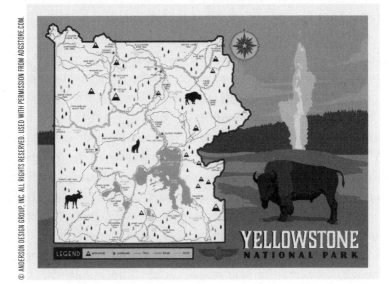

world's largest dark-sky preserve (zero light pollution). The other, Nahanni, is close to a 4 Y and is about as spectacular as any park could hope to be. Voted one of the Seven Wonders of Canada, Nahanni has jagged peaks called the Vampire Spires; a collection of mountains and sheer rock walls known as the Cirque of the Unclimbables; and the waterfall that dwarfs Niagara. Throughout the Boreal, there are something like 2 CALs of forest now being considered for protection. Plus a CAL in Alaska, where 25 percent is off-limits to development, the closest thing the United States has to half of a Half Earth landscape.

Carbon bomb

The Boreal Forest will never be farmland—that's not a threat; its soils are too thin and too acidic and the growing season too short, and when it's warm there are too many insects, and when it's cold you can't raise crops under the northern lights. Clear-cut timbering has subtracted 14 Ys (28 million acres) of old-growth trees in the last twenty years, and the forest also has belowground riches that could be pumped out or mined, such as oil and gas, diamonds and gold. And, perhaps more precious than any of these, are deep layers of peat beneath the wetlands that give the Boreal the nickname "the Fort Knox of carbon."

The Boreal, like all landscapes, is vulnerable to global warming. It's estimated that each decade the tropics expand their range by thirty miles, while northern regions are heating up twice as fast as the rest of the planet. But the Boreal is also a vault, a sink, a storehouse, a reservoir, a bank (terms used interchangeably)— meaning a place that sucks carbon from the air and tucks it away in tree trunks and peat bogs, hanging on to it instead of spewing ever greater amounts of planet-warming carbon dioxide into the

atmosphere. Disappearance of the Boreal, it's said, would deto-
nate a carbon bomb.

This whole absorption process is largely invisible, and one
aim is to keep it hidden. Just as anti-fossil-fuel activists who
oppose new oil wells have rallied round the cry "Keep it in the
ground," one slogan for the Boreal could be "Keep the ground in
the ground." It's still not appreciated how the Boreal locks away
twice as much carbon per acre as tropical forests. "The Carbon
the World Forgot" is the title of a 2009 report on the subject.
Aboveground, tropical trees hold as much carbon as Boreal trees,
but belowground the Boreal has been amassing carbon since the
last Ice Age, and adds to it every year. Peat, a singular spongy
substance that Jeremy Leon Hance, an environmental writer,
describes as having "the color and texture of moist chocolate
cake," is the key to all this.

In the forest of blue, mosses and other plants growing next
to pools originally left behind by retreating glaciers fall into the
water at the end of the growing season, but the acidic water doesn't
let them fully decompose or release the carbon taken from the air.
Instead, their presence continues under water, and over time this
ever more compressed layer of summers past becomes peat, with
ancient plants at the bottom and more recent ones up top, and as
it compacts it gets pushed deeper and deeper underground. After
eight thousand years of this slow accumulation, Boreal peatlands
can extend downward nearly 30 feet. In the upper Mackenzie
River Valley, there's soil as far as 330 feet below the surface that
is tightly gripped by permafrost (tropical soils, by contrast, even
when several feet thick, have little organic matter). The Boreal
has 3 CALs, about 294 million acres, of peatlands, one-third of
the world's total.

"Carbon bomb" refers to the fact that, if the Boreal were felled
and the peatlands drained, this would release as much carbon

dioxide into the air as all the fossil fuels humanity has burned over the last thirty-six years. The Boreal is a global braking system, extending perhaps until midcentury the time still left to keep climate change in bounds.

Moose, wolves, and caribou

As warming reaches the Boreal, the expectation is that its enormousness can help accommodate or blunt any impacts. So far, one of the most noticeable changes is all those moose. Moose have been moving north as the shrubs they eat move north (as pointed out in a 2016 study by five Arctic scientists). This could be one factor interrupting what had been the ebb and flow in the numbers of the Boreal's (and Canada's) most iconic animal, the caribou—or reindeer, as they're known in Scandinavia and Russia, where they're herded and used to draw sleds (the inspiration for Santa Claus on Christmas Eve). How important are wild caribou to Canadians? "Think America's bald eagle," Josh Axelrod of the Natural Resources Defense Council suggests in a blog post, "or China's giant panda."

A caribou has been on the Canadian quarter for most of a century; in profile, showing velvet-covered, flattened antlers, it's an elegant symbol of what one hundred years ago was both ubiquity and bountifulness. Over thousands of years, different groups of caribou adapted to forest life or life in the mountains, or to the treeless tundra north of the Boreal, where thousands of these so-called barren-ground caribou make annual 2,500-mile migrations, one of earth's great wildlife spectacles. Still other caribou find a home in even harsher conditions amid the ice fields and bare rock on remote High Arctic islands.

Wolves eat caribou, and in the Boreal wolf and caribou popu-

lations have always fluctuated in tandem, often dramatically so. When there are a lot of caribou (with peak populations in the hundreds of thousands) more wolves can feed, but then the increase in wolves reduces the number of caribou to the point where the wolves themselves die back precipitously. After that, over a twenty- to thirty-year period, the cycle repeats. But now that's not happening, and according to a 2017 study by British Columbia and Alberta research ecologists (titled "Experimental Moose Reduction Lowers Wolf Density and Stops Decline of Endangered Caribou"), the arriving moose are interrupting the pattern—the elephants in the room, you could say.

Both moose and caribou are deer, and so cousins of a sort, but moose can be bigger than horses and weigh four times as much as caribou, making them that much harder for wolves to catch. But they're not uncatchable, even the adults, and now wherever moose arrive, Boreal wolves have an alternate food source. Which means that wolf numbers won't automatically crash when caribou numbers are at a cyclical low (in some cases a 90 percent, even a 99 percent, decline). And wolf numbers that remain high make it that much harder for remnant caribou populations to rebound.

Caribou are unusually resilient. Their hooves have more func-
tions than a Swiss Army knife, acting like paddles when swimming across a river, like snowshoes in winter, and also like a scoop
to dig under snow to find lichen, twigs, leaves, and moss to eat.
But caribou are also unusually helpless in the face of changes they
haven't experienced before—for instance, in their daily roaming, which can take them fifty or sixty miles, woodland caribou
will avoid any area regrowing after a forest fire for fifty years
or more. Since fires come and go repeatedly throughout a forest, this means woodland caribou need something truly big to
live in, a forest most of which won't have burned for at least half
a century.

Moose arrival has only compounded things for these shy
and sensitive creatures. Because of clear-cuts and development
in southern Canada (and the northernmost U.S. states), total
caribou habitat is now half what it once was. In 2012 the Canadian government gave the provinces and territories five years to
develop caribou protection plans. Not one met the deadline.

Guardians

The government could have turned to the Indigenous community
for guidance. Valérie Courtois is director of the Indigenous Leadership Initiative, a group that strengthens Indigenous efforts to
reclaim and conserve their land (Stephen Kakfwi's been a senior
advisor). The best plan for the resurgence of caribou, she says, is
to give Indigenous people responsibility for the job. "No question," she told me on a visit to New York, "it will lead to better land
management, since the Innu, for instance, have been thinking
about this for nearly ten thousand years."

Valérie Courtois describes herself as half Innu. Her mother's
Québécoise and her Innu father was a Mountie, a Royal Canadian

Mounted Police officer. She is young, has boundless energy, and is constantly on the move, explaining, encouraging, exploring possibilities, celebrating successes. She's a forester, a planner, and a photographer—someone equally at home in what she refers to as Innu science and the principles of Western science (conservation biology). Which is helpful because government scientists are comfortable thinking twenty-five years ahead, while for Innu planners that's like calling an appetizer a meal or a consultation a cure. The minimum time frame they're willing to consider is four hundred years. For them, there's a doubled urgency—their own survival as well as that of the caribou. The Innu are, have been, will always be the caribou people.

There's a story Courtois first heard from her grandparents, one of those Indigenous stories that goes to the heart of things. An Innu boy, out hunting with his father, both of them starving, had a dream the night before that he has married a caribou and is no longer hungry. Seeing a pond and caribou tracks all around, the boy draws a bow, but one of the caribou looks him straight in the face and he can't shoot. The caribou, which can talk, proposes marriage. The boy says he can't move across the snow the way caribou can, but is told he'll be able to follow. The boy says he won't have his tent, but is told he'll be warm. The caribou then transforms into a beautiful maiden, and the boy falls in love. There's a ceremony, a sacred marriage, and for the rest of his days he sees her as a maiden. But ever afterward she sees him as a caribou.

"This is the sacred agreement between the Innu people and the caribou," Courtois says, "and from it flow all our ethics and rules. We're married to the land. Our whole lifestyle revolves around following migratory herds, so we, like them, need a lot of space to be who we are as a people. Our elders say that if the caribou disappear, we won't be Innu anymore. In the planning that gets taught in schools, first you mark off areas for industrial devel-

Valérie Courtois

opment, then you decide what to protect. The government has suggested a twelve percent set-aside. But in the planning that gets taught by the land, we start out asking what needs to stay where it is for Indigenous people to stay who they are. What will you need to remain you? Indigenous thinking leads to Half Earth figures, or higher."

For Smithsonian anthropologist Stephen Loring, who's done fieldwork with the Innu in Labrador, the story of the dreaming boy doesn't go back far enough in time, as he explains in his essay "At Home in the Wilderness": "The antiquity of the relationship between human beings and caribou extends back to the Ice Ages. It is not improbable that we became human because of caribou: that core human traits such as cooperation, language, and social identity were first forged, or certainly reinforced, around Pleistocene campfires in both the Old World and the New."

The George River caribou herd in Labrador, a migratory

herd that divides its time between tundra and forest, is the one that's seen the steepest drop in numbers. Courtois remembers her first drive across the province, how she had to stop for an hour while the herd crossed the road. Back in 1993, there were 800,000 caribou; in 2017, 5,500. In 2013, for the first time, seven Aboriginal nations representing Inuit, Cree, Naskapi, and Innu, as well as Métis, worked together, creating UPCART, the Ungava Peninsula Caribou Aboriginal Round Table. And after four years of meetings, they produced a plan of their own, called "A Long Time Ago in the Future," a fifty-seven-page caribou strategy for a truly immense piece of land, 370 million acres (about 4 CALs) in Labrador, Quebec, and Nunavut. Courtois recalls an Innu elder saying, "Our unity is a gift from the caribou. We have to honor it."

This is just one of the unpredictable effects of global warming—no one had anticipated that caribou numbers could be suppressed by moose. Equally, no one expected a breakthrough alliance among age-old champions of the caribou.

The plan: Indigenous Guardians. Innu, Métis, and Cree, trained by the elders and by biologists, would fan out across this cold northern landscape, said to be the spot from which glaciers once spread south as far as New York City. These people, hired like park rangers, would take on a number of jobs: inventorying caribou herds; monitoring calving ranges and the refuges caribou retreat to in times of scarcity; and protecting them from things like low-flying helicopters, chartered by mining companies that extract nickel, gold, and iron ore. Caribou hate the noise so much they take off as fast as they can, and their top speed is fifty miles an hour. Often they fall or exhaust themselves.

For Valérie Courtois, Indigenous Guardians are, instead of boots, "moccasins and mukluks" on the ground. "What are guardians?" she asked a Parliamentary committee in 2016. "They're essentially the eyes and ears of communities on the lands. I directed a program for almost a decade with the Innu nation in Labrador, and our favorite saying was, 'Today's guardians are tomorrow's ministers.'" She told me that "these future government leaders, like the rest of us, are looking for the Boreal to be not only the best-conserved but the best-managed terrestrial ecosystem in the world."

As in the story of the Innu boy, the stirrings of the idea for Indigenous Guardians began with a dream—a dream that Captain Gold, from the Haida First Nation in British Columbia, had in 1973 about the Haida Gwaii. These islands, called the Canadian Galápagos, are a 155-mile-long archipelago of more than two hundred islands 30 miles from the British Columbia shore, where rain-drenched, windswept, moss-draped red cedar and Sitka spruce forests with towering trees spill down rugged peaks as high as 3,800 feet and reach almost to the water's edge.

Haida art is similarly towering, particularly their astonishing skyscraping totem poles, more properly called "house frontal poles," up to one hundred feet tall and lined up in a row along the edges of waterfront villages, designed to be seen from canoes offshore. Each pole is carved from a single giant red cedar tree trunk into a distinctive, vertically stacked family crest made up of heraldic figures with large, bold eyes, one on top of another—eagles, ravens, bears. Once there were hundreds of these waterfront villages, but in the nineteenth century smallpox and other epidemics reduced the Haida population by 90 percent.

Captain Gold (a name he adopted to honor an ancestor) grew up in Skidegate, the small town to which the remaining Haida population had relocated. He kept hearing about an abandoned village, the southernmost one, SGang Gwaay, once his family's home. This place had great spiritual strength, he was told. He started dreaming about it. He had no idea how to canoe but decided to buy a sixteen-foot red fiberglass canoe from a Sears catalog, and after fitting up a sail and a makeshift outrigger, paddled by himself for one hundred miles to his destination.

No one had lived in SGang Gwaay for almost a century, but as he drew near he found he could see it as it used to be, with children diving and swimming beside him. But he could also see what it had become—the poles, some of them two hundred years old, were in danger of toppling, and the area was overgrown with brush and strewn with debris from picnickers and pillagers, specifically "pot hunters," who dug for artifacts. Gold took it upon himself to honor the ancestors by clearing the village and taking care of the place. He would spend his next twenty summers there—and once he got to work, things, as he put it, snowballed.

Watchmen—villagers in tall, red-and-black-striped conical hats who warned of danger—had always been part of Haida communities. Their images can be found on the top of many house frontal poles. Captain Gold recruited volunteers to set up base camps at other abandoned villages to clear brush, give tours, and control visitors who wanted to climb the poles. It was his idea that the volunteers collectively take the name Haida Gwaii Watchmen. There are now about twenty-five Watchmen (and women) throughout the area every May to September.

In 1985, TV cameras recorded seventy-two Haida elders, in their ceremonial black-and-red blankets with mother-of-pearl decorations, when they got arrested and handcuffed for blocking a logging road—clear-cutting timber companies were getting too close. In that moment, protecting Haida Gwaii became a national cause. In 1991, SGang Gwaay became a UNESCO World Heritage Site "of importance to the history of mankind."

The southern third of the island chain has since been designated the Gwaii Haanas National Park Reserve and Haida Heritage Site, co-managed by the Haida Nation and by Canada's

Traditional image of Haida Gwaii Watchmen

park agency. More recently the site has been named a National Marine Conservation Area Reserve as well—so that it now includes offshore waters, feeding grounds for blue and gray and humpback whales, porpoises, dolphins, seals, sea lions, orcas, and millions of seabirds. Unlike any other protected landscape in the world, this million-and-a-quarter-acre park extends without interruption from mountaintops down to the ocean floor, eight thousand feet below sea level. The whole process of rescue and reclamation took thirty-seven years and involved the work of hundreds of people—beginning with a dream that prompted the purchase of a red canoe.

The Haida Watchmen have inspired about thirty Indigenous Guardian programs around Canada, but what hasn't happened yet is what's happened in Australia. There, in the years since the High Court abandoned *terra nullius* and restored native title to about a third of the country, Australia has created a National Reserve System of parks that covers almost a fifth of the country and incorporates an ever-growing number of Indigenous Protected Areas—seventy-five at last count on 165 million acres, including one with 25 million acres of northern grassland and desert, making it the biggest reserve on the continent.

The Australian government has spent $680 million to fund Working on Country, a program that's hired more than two thousand Indigenous Rangers (their Indigenous Guardians) to maintain the environmental integrity of the 165 million acres. This involves tackling some of the most pressing ecological problems, particularly out in areas that, after the Aboriginals got displaced, were overrun by animals introduced from elsewhere and then abandoned—cane toads, rabbits, even camels (Australia has more than a million one-hump and two-hump camels, the largest feral camel herds in the world, and the number could double in a decade).

"Working on country," as opposed to "in the countryside," is more than a local quirk or semantic distinction; the phrase has special meaning for Aboriginal Australians, a way of indicating that the land is alive and lived with, not in. "People say that country knows, hears, smells, takes notice, takes care, is sorry or happy," Deborah Bird Rose, an ethnographer who's spent years working with Aboriginal people, says in her book *Nourishing Terrains*. "Country is home, and peace; nourishment for body, mind and spirit; heart's ease."

Canada is racing to catch up with Australia—or maybe with itself. "This is Canada's crazy moment," Harvey Locke, a Canadian lawyer (no longer practicing) and a conservationist of large landscapes, explained to me. "A fraught and exciting time for a country that no longer sees itself as a son of the British Empire or the younger brother of the United States."

A central issue in the campaign that made Justin Trudeau prime minister in 2015 was reconciliation with Indigenous people. Once in office, Trudeau embraced "Canada Target 1," a commitment to protect 17 percent of its land (and 10 percent of its oceans) by 2020. This was actually a recommitment to an earlier international pledge—one of the Aichi Biodiversity Targets, agreed to in 2010. But in a country where it can take five years to approve a single new park, this sounded outrageously ambitious because of the amount of land involved and the speed needed. In 2015, 10.6 percent of the country was protected; by 2020, only 12.5 percent. So—another 111 million acres to reach Target 1.

Putting some wins on the board

This was the "big chance" Steve Kakfwi and Steve Kallick had been looking forward to for years. Steve Kallick likes to think of himself as a practical person—"practitioner" is a word he prefers, not visionary.

Back in 1997, Kallick was with the Pew Charitable Trusts, in Philadelphia, beneficiary of much of the fortune of Joseph N. Pew, founder of the Sun Oil Company, when he and his colleagues began thinking about the Boreal. They were also working on the roadless rule, in what Kallick's boss at the time, Joshua Reichert, who has a Princeton doctorate in social anthropology, called "a results-oriented way." In three years, the roadless-area campaign permanently protected 58.5 million acres of wilderness in U.S. national forests, by pulling what Kallick calls "an aikido move": the Forest Service had been asking the public how to manage its existing network of roads, and the campaign turned that around into a referendum on whether there should be any more roads in the remote backwoods at all. This generated 2 million "no more roads" public comments, along with six hundred anti-road editorials and thousands of newspaper articles.

One of the things Kallick likes about working in Canada is that it, like Australia and the United States, has money, land that's still protectable, a history of conservation, and operates as a rule-of-law country, meaning that decisions are likely to stick and not get arbitrarily overturned. These are all circumstances that, he said, make it easier to keep thinking positively by putting some wins on the board. One question, early on, was how much of the Boreal to protect—this at a time when 12 percent seemed daring. Kallick remembers an informal meeting at a ski lodge in Sutton, Quebec, in which Harvey Locke set off clouds of chalk dust by thumping on a blackboard (it was that long ago), imploring him to "ask for what we fucking need"—at least 50 percent. "You know what?" Kallick said. "He's right. Let's be bold!"

The Canadian Boreal Forest Conservation Framework, announced in 2003, seemed wild-eyed at first, even to environmentalists—"So far outside the mainstream," Steve Kallick remembers. But some big timber companies said it sounded reasonable, and First Nations groups signed on too, even though to them the idea sounded skimpy. The proposal called for full protection of 50 percent of the Boreal in connected reserves and allowed for development in the other 50 percent—but only by loggers, miners, and hydro-dam builders who respected the ecosystems.

Less than a decade later, the framework's fifty-fifty split seemed practically conventional: after surfacing as a last-minute, hot-button issue in a 2010 Ontario election (Ontario is the most populous province), the government there passed a Far North Act to protect half of the top of the province, about 55 million acres of the Boreal, and a year later neighboring Quebec (Quebec's the largest province) adopted a Plan Nord to make nearly 150 million acres—half of its own north—off-limits to development. Together these two provinces are home to three-fifths of Canadians and are considered the country's core region, its historical hearth and heartland.

But what happens when the second 50 percent, where development can take place, has first 50 percent qualities that can't be compromised? In the Hudson Bay Lowlands in Ontario's Far North, the two 50s are actually stacked on top of each other. The Lowlands are part of the forest of blue, home to ten thousand First Nations inhabitants and teeming with wildlife, including polar bears, millions of snow geese, and, yes, mosquitoes—as many as 5 million per acre. The ground beneath, ten thousand years old, is a "crown jewel," says a booklet called *10 Cool Canadian Biodiversity Hotspots*, "the single largest peatland system" that's "potentially the most carbon-dense terrestrial ecoregion on Earth." But way below the peat, almost five hundred feet down, and spread over more than a million acres, is the Ring of Fire, named by a mining promoter for the Johnny Cash ballad, a crescent-shaped rock formation so dense with nickel and chromite (used to produce chromium and stainless steel) it could be worth $60 billion and support mining operations for a century.

The quieter part of the Boreal Forest Conservation Framework strategy—the funding of First Nations groups to undertake conservation on their own—may do more to secure the Boreal's future than anything else. There are already two dozen Indigenous Protected Area proposals under consideration, and these actions are giving Canada a new way of looking at itself. The Edéhzhíe Protected Area, for instance, in the Mackenzie River Valley, the first IPA paid for by a $1.3 billion fund set up as part of the Pathway to Canada Target 1, excludes mining and drilling from a 3.5-million-acre plateau considered a "breadbasket" by the local Dene, since it's never deserted by wildlife, even in times of scarcity. (The local Dene are known as Dehcho Dene—"Dehcho" means "big river," the Mackenzie.)

The plateau also sits on extensive lead and zinc deposits, and possibly also on oil and gas reserves, and a belt of diamonds one hundred miles thick. The pre-Trudeau Canadian government, after years of negotiation, had no objections to seeing the surface protected but sought to make sure that anything underground would still be extractable. Edéhzhíe is now doubly protected. It's also been designated a National Wildlife Area and will be staffed by Indigenous Guardians and by the Canadian Wildlife Service. And according to the Trudeau government, it's the first of many IPAs that will advance reconciliation.

Water heart

Déline is one of those Mackenzie River Valley villages that can only be flown into. To get there, you fly over Great Bear Lake, eighth-largest lake in the world. Looking down you may see what look like logs lying on the rocky lake bottom. But they're not logs, they're fish—the giant lake trout that can live sixty years and weigh eighty pounds. The water in Great Bear Lake is that clear and pure. Déline is the only settlement along the nearly 8-million-acre lake, which has five arms (or two arms, two legs, and a head, if you're thinking bear shape), a shoreline that's 1,700 miles long, and is so vast it makes its own weather. Déline's basically a single lakefront street of cottages and bungalows for about five hundred inhabitants. From above, you can see the town's twin magnetic poles, southern and northern, so to speak.

A tall white wooden cross at one end marks the site where an Arctic explorer, Sir John Franklin, wintered in 1825. Southern Canada knows about Déline because close by is a small lake considered the birthplace of ice hockey—Franklin's men invented the game one day on its frozen surface. Until 1993 the name of the town was Fort Franklin. The very last house at the other end of the street, a log cabin with a blue roof, stands by itself at the edge of a stream with only open country beyond. This is the prophet's house, a place where people gather for contemplation and stories, a place to discuss the sayings of Eht'se (Grandfather) Louis Ayah.

In 1865, when Ayah was eight, he began having visions that offered him glimpses of the future, some of them intriguing, others harrowing. Ayah died in 1940. Because of Ayah's encouragement and forebodings, Déline, after twelve years of negotiation, became, in 2016, the first self-governing Aboriginal community in the Northwest Territories. That same year, 23 million acres of the Boreal, including the lake, were named the Tsá Tué Biosphere Reserve—the largest in North America and the first to be Indigenous led, by the Sahtúot'ine, the Great Bear Lake Dene. The village constitution says, "We are the descendants of the prophet Eht'se Ayah."

It was said of Louis Ayah that he knew you were coming to visit before you got there, and as soon as you arrived, he knew what was troubling you. Once he pointed to an area beyond the lake where rock that "will look like glass" would someday be discovered, and

A giant lake trout from Great Bear Lake

decades later diamonds were found there. His son-in-law related that another time, out with hunters, Ayah saw in his mind a place 165 miles from Déline, on another arm of the lake, that would be filled with houses, people with white skin, tunnels in the ground, big boats, and a plane. "Because I have Indian medicine," Ayah said, "I thought they wouldn't see me, so I went with them on the plane." He saw the plane heading south, transporting rock that was "dangerous stuff" to be made into weapons and then put on another plane. Again Ayah went with the plane, which flew over the ocean and dropped a bomb on people who, as the son-in-law recounted, "looked more like us." It's a Sahtúot'ine story set down in *If Only We Had Known*, a handsome book published in 2005 by the Déline First Nation.

What Ayah was seeing was Port Radium, now dismantled, where radium would be mined in the 1930s, and silver in the 1960s and '70s. During World War II, it was one of Canada's most secret military sites, a uranium mine that supplied ore to the Manhattan Project for the bombs dropped on Hiroshima and Nagasaki. Dene porters, who had not been warned about the hazards of radiation, were paid three dollars a day to work twelve-hour shifts, carrying uranium ore in burlap sacks. Many later developed cancer, and Port Radium became known as the Village of Widows. In 1998 a group of Dene went to Japan to apologize to those who had survived the attacks.

Ayah's most ominous prophecy was that very close to the end of time, famine will stalk the earth and everything will dry up, and people will come to Great Bear Lake, the last refuge, in search of fish and water. Holding his fingers together, he told people, "That is how many boats will come fill up the lake"—so many that someone could cross it without getting wet, simply by walking from deck to deck. The prophet's house is a single large, comfortable room ready for gatherings, with sofas and benches lined up along the walls. I sat there with Danny Gaudet, an outgoing, voluble Dene businessman who'd been chief negotiator for Déline self-rule, and asked him if he considered this prophecy inevitable. Outside, an eagle perched on a power line—a good sign, Gaudet said, glancing around the room. As for the prophecy, he thinks that remains to be seen. It's not enough if only governments and industry are present—"You need one more player." Indigenous people constitute only 5 percent of the planet's population but they dwell on a quarter of the land. Now that they're finding a voice and being listened to, there's a chance to get the balance right, a chance for the world to be helped on its journey. Christopher Filardi, an American conservation biologist who's visited Déline, is more confident. Indigenous people, he told me, "aren't just the past—they're the future of the world." Another prophecy from Ayah, retold by Morris Neyelle, a Déline elder: "You will get powers," Ayah said, raising his hand high over his head. "That's how far you will rise."

In the quiet of the prophet's house, my sense was that any number of things—the Boreal campaign, reconciliation, IPAs, Indigenous Guardians—might be happening only just in time. Had the residential schools lasted another generation, for instance, those last-generation activists might never have banded together. Steve Kakfwi lost all his Dene language in residential school and had to relearn it before he could rally people against the Mackenzie Valley Pipeline. Part of the therapy he devised for himself was lying on his back on a hill in Fort Good Hope, listening to what the land could tell him.

Later that day—my last in the North—Bruce Kenny took me out on his boat so I could taste Great Bear Lake water for myself. He's a prodigious, high-spirited fisherman who also drives the Zamboni machine between periods in local hockey games and plays in two hockey leagues. A Sahtu Dene story from back before Louis Ayah's time talks about Tudzé, a water heart that beats at the bottom of the lake, protected by the fish, alive itself and connecting all other living things. Five miles from town, I watched Kenny land a twelve-pound lake trout, not enormous but big enough to take back to the elders. Then he dipped a cup into the lake and offered it to me. The water tasted exactly as it must have tasted since the glaciers receded, fresh and cold and uncontaminated.

When I glanced up, I was suspended between a beating heart below and an enormous Group of Seven sunset canvas in the sky. There was an intense golden glow along the bottom edge of a dark band of low clouds near the horizon, throwing glints onto the water. Above the clouds, the blue-white sky turned orange and then became bright salmon stripes that seemed to keep reaching higher and higher. I had to lean way back to see over the top.

All of Life Inside and Nowhere Else

The Boreal, caribou, humanity; the continents, the oceans, the earth's crust and atmosphere. All these are interlocking aspects of a single ancient and encompassing realm, the biosphere. Partly invisible, remarkably crowded, quirky, bizarrely shaped, with much of it still unknown, the biosphere sustains all of life—and gets taken for granted. Setting aside half the earth involves recognizing that the biosphere is home, our frame of reference, the complex context for everything.

Of all the biosphere's features, perhaps the most astonishing is its thinness. A letter's envelope, an apple's skin, a turtle's shell, or a lobster's—these are outsides, coverings, wrappers. But what about when an envelope, paper-thin as it is, is the letter itself, content as well as container, like the featherweight fold-over European "aerogrammes" so popular during the austerity days after World War II? Or where a skin holds the nourishment of the fruit, or a shell animates an animal, its outwardness doubling as inwardness? What if, when talking about the earth, this isn't a hypothetical?

Vladimir Ivanovich Vernadsky

A century ago, Vladimir Ivanovich Vernadsky, a Russian bio-geochemist, showed that to be the case and introduced us to where we live in a then-obscure book called *Biosfera* (initial press run: two thousand copies). The biosphere, as he described it, is at once solid, liquid, and gas, part land, part water, and part air. It burrows into the earth but also hovers over it, forming a zone and layer of life confined to, as Vernadsky noted, "the surface that separates the planet from the cosmic medium."

A good deal has been learned about the biosphere since *Biosfera* appeared in 1926—it has been called an indispensable term of modern scientific understanding, and across the planet citizens and governments have now protected 701 so-called biosphere reserves in 124 countries. The pace of discoveries about the biosphere is accelerating, with excitement to match. "This is biology's golden age," Dr. Victoria Elizabeth Foe, a developmental biologist, said in *The New York Times*. She sees the exploration of the biosphere as analogous to the cathedral building of a thousand years ago: "Some of us are building arches, some painting murals, some carving in stone." Ed Turner, curator of insects at the University of Cambridge's Museum of Zoology, has told an interviewer, "It's an incredibly undiscovered world."

A couple of English earth scientists, Tim Lenton and Andrew Watson, begin *Revolutions That Made the Earth*, a book about the entangled history of life and the planet, by telling humanity, "You are slowly awakening from the deepest of sleeps. Your sleep has lasted since the beginning of time. Dreams still echo in your head, myths of creation, voices, music, fragments of beauty and terror."

Vaclav Smil, a geographer considered the world's foremost thinker on energy of all kinds, begins his 2002 book, *The Earth's Biosphere: Evolution, Dynamics, and Change*, by quoting the first-century Roman philosopher Seneca the Younger: "The day will come when posterity will be amazed that we remained ignorant of matters that will to them seem so plain." These days, amazed and alarmed. Smil's books are getting some extra attention now because he's one of Bill Gates's favorite authors. Gates has read nearly all of Smil's thirty-seven books, and says, "I wait for new Smil books the way some people wait for the next Star Wars movie."

Life is everywhere and nowhere, as *Biosfera* describes it—it's everywhere inside the biosphere, and nowhere else, anywhere. The biosphere is permeated with life because it has, according to Vernadsky, "tended to take possession of, and utilize, all possible space," to the point where it's now "spread over the entire surface of the Earth in a manner analogous to a gas."

How much life does this global cloud contain? The head count actually began two hundred years before *Biosfera* was published, when Carl Linnaeus, a Swedish physician and *Princeps Botani-*

corum (the Prince of Botanists, his friends called him), famously began inventing and affixing Latin first and last names to European plants and animals, the genus and species kids learn about in grade school, both names italicized and always with a capital letter for the first name. By the time Linnaeus's ever-expanding roster of names—a kind of telephone book or card catalog of life—reached its tenth edition in 1758, it included 4,400 animals and 7,700 plants. "God created, Linnaeus organized," he said of himself.

Seventeen of Linnaeus's devoted students—he thought of them as "apostles"—began collecting and naming species and applying his methods in North America (Pennsylvania, New York, New Jersey, and Canada) and countries as far away as South Africa, China, and Japan. The Linnaean system is still the standard the world over, and at this point 1.9 million species have been identified and named. The two best-known names in the system, it's been said, actually postdate Linnaeus—*Homo sapiens* and *Tyrannosaurus rex*.

Eriovixia gryffindori, *the Sorting Hat spider*

Estimates of how many more species there are vary from 8.7 million to perhaps a trillion. Even if 8.7 million is right, we're only a fifth of the way there. In one recent year, 18,000 new species were named, and that's a typical annual figure. One of those 18,000 was a single sample of an Indian spider, found in the Western Ghat mountain range. It's called *Eriovixia gryffindori* because, according to the *Indian Journal of Arachnology*, its tiny, cone-shaped body looks like the Hogwarts Sorting Hat. The people who named it hope this will stand "for magic lost, and found" and "draw attention to the fascinating, but oft overlooked world of invertebrates, and their secret lives."

Rivers and arteries

The *where* of life is also expanding, with discoveries like the "deep, hot biosphere," whose existence in the earth's crust was first talked about in a 1992 paper by Thomas Gold, a Cornell astrophysicist. This realm is the true rock bottom of the food chain, Hubert Staudigel, an oceanographer who studies rock-eating microbes, explained to the BBC. Many deep biosphere bacteria have no need for the sun's energy or oxygen or carbon dioxide and are being found a mile and a half beneath the floor of the sea. And maybe down deeper than that, once deeper holes get sunk farther into the planetary crust. Some individual microbes metabolize so slowly they've been dubbed Methuselah microbes and could be millions of years old. Tom Davies, from the Texas A&M Ocean Drilling Program, told *Science Daily* that exploring the deep biosphere is "like walking into a tropical rainforest for the first time and beginning to identify and count the birds."

The biosphere is filled with unexpected, unsensed oddities, such as intermittent invisible rivers. Over the oceans, a half mile up or so, maybe a little higher, are something like eleven atmospheric rivers (ARs) at any one time, two-thousand-mile-long corridors of fast-moving, moisture-laden air sometimes called "horizontal hurricanes," carrying as much water as the Amazon. The most famous one, the Pineapple Express, periodically dumps water on the California coast, water carried from near Hawaii.

ARs are considered an emergent phenomenon, which means we haven't known about them for long and their origin is unknown, though NASA predicts that by the end of the century these rivers in the sky will get wider, longer, and stronger. There are forest-based ARs as well—"Every tree in the forest is a fountain, sucking water out of the ground through its roots and releasing water vapor into the atmosphere through pores in its foliage," says an article published by *Yale Environment 360*. An AR that originates over the Amazon sends rainfall three thousand miles away to the cornfields of the U.S. Midwest—and may begin to dry up as more and more of the rainforest gets felled.

One hundred and fifty feet below Toronto flows an iron-rich, slow-moving, cold, and drinkable river, the Laurentian. It's maybe 5 million years old. It seems to stem from Georgian Bay, an arm of Lake Huron about seventy miles north, and may continue to flow under the floor of Lake Ontario. It's at the bottom of a valley in the bedrock long since buried by glacial debris, and it dramatically announced its presence in the city in 2003 when, during renovations to the city's largest park, a just-dug well exploded, sending water and gravel the size of golf balls fifty feet into the air, the *Toronto Star* reported.

Periodically, giant currents of mud and sand almost three hundred feet tall pour down the steep slopes of Monterey Canyon in the Pacific Ocean below Monterey Bay. The canyon has walls a mile high (or deep)—which makes it Grand Canyon height. This intense underwater flow, which plummets much like an avalanche, is one of an uncounted number of undersea rivers called the arteries of our planet because they bring oxygen and nutrients across the deep ocean floor. A 2017 BBC article by Richard Gray quotes a British sedimentologist, Dan Parsons, who calls undersea rivers "incredibly powerful and destructive flows." They remain, Gray says, "among the least understood phenomena on our planet."

Virtually unknown until the 1980s, some of these undersea (and unnamed) rivers are thousands of miles long. Some are straight, some sinuous. Many extend the flow of enormous and long-since-named rivers like the Amazon and the Congo, but others are right out in the middle of oceans with no discernible source. An undersea river along the bottom of the Black Sea, a mile below the water's surface, carries dense salty water from the Mediterranean that's heavier than Black Sea water—so much water that, if it were on land, it would be the sixth-largest river. Over the last six thousand years, it's carved a channel 115 feet deep, creating underwater rapids and waterfalls.

The geological force

So there's what's known about the biosphere and what's yet to be known—an amalgam of golden age and slow awakening. This could also describe the international reputation of Vladimir Vernadsky, who died in 1945, at eighty-one. In addition to *Biosfera*, he wrote 415 books and papers on a wealth of topics. He labored in considerable obscurity in Russia under Stalin. Since then he's been lionized there and in Ukraine, where he's considered an Einstein or a Darwin. His image—a sweep of white hair, bushy snow-white beard and mustache, round wire-rimmed glasses, thoughtful frown, and an abstracted gaze into the deep distance—

has been on postage stamps. A Moscow Metro station, volcanic mountains on a Russian island, a peninsula in Antarctica, and a crater on the dark side of the moon—all have been named in his honor. But in the English-speaking world and the West in general, well . . . "Ever heard of Vladimir Vernadsky?" asked a *New Scientist* writer, Fred Pearce, some years ago. "Probably not."

Considered a founder of several crossover fields, such as biogeochemistry and radiogeology, Vernadsky ignored boundaries. The central insight of *Biosfera* was foreshadowed, Smil notes, back before the Russian Revolution when Vernadsky jotted down the question he then spent years working out: "What importance has the whole organic world in the general scheme of chemical reactions on the Earth?" Before Vernadsky, contemporary biologists were either examining life with no reference to the environment or seeing life as passively adapting to the geological conditions of the time. *Biosfera* proclaimed the two inextricably linked. Inside the biosphere, life shapes geology, or, as the 1998 foreword to *The Biosphere*—the belated full English translation of *Biosfera*—put it, life is not merely *a* geological force but *the* geological force. The authors of *Revolutions That Made the Earth* say that what scientists do now is "weave many strands of science together," a perspective that "considers the evolution of life and of the non-living environment as one coupled, indivisible process."

Connection is not a metaphor; it's an underlying principle, the essence of it. The biosphere's inhabitants don't simply dwell in it, they shape its parts—whether solid, liquid, or gas—again and again, often profoundly. The inhabitants are then shaped by the new configurations of land, water, and air that they themselves have brought about. Cause and effect, effect and cause.

In the early days of life, perhaps 3.5 billion years ago, the oceans were yellow green and the sky most likely orange. Then came red seas, as cyanobacteria began to accumulate there. The oceans and the sky turned blue over several billion years as they filled with oxygen generated by more bacteria—this was the "third earth," a phrase used by Smithsonian paleobiologist Douglas Erwin. More recently, about 400 million years ago, the next color arrived—life moved ashore and the land began to green.

Burst the limits of time

More than a century before *Biosfera,* the earth started getting old; for a long time European scientists had thought it to be youthful. Reasoning from the Bible and from what was then known about ancient history, a seventeenth-century Irish archbishop, James Ussher, calculated that the world was created precisely on October 23, 4004 BC.

In June 1788, the Scottish geologist James Hutton, now revered as the father of modern geology, took friends on a boat to Siccar Point east of Edinburgh to show them the exposed rocks of this spectacular steep seacoast cliff; these days it's a place of geological pilgrimage. He pointed out the juxtaposition of great horizontal slabs of red sandstone sitting directly on top of tall, practically vertical layers of greywacke sandstone. This could only be explained, he said, by millions of years of uplift, erosion, and tilt, and an ocean that came and went. One of Hutton's friends, the Reverend John Playfair, wrote, "On us who saw these phenomena for the first time, the impression made will not easily be forgotten. . . . The mind seemed to grow giddy by looking so far into the abyss of time."

A generation later, Sir Charles Lyell achieved fame popularizing Hutton's ideas (Lyell's buried in Westminster Abbey and, like Vernadsky, has a crater on the moon named for him). Lyell found the same "ideas of sublimity" in Hutton's discoveries: "The imagination was first fatigued and overpowered by endeavoring to conceive the immensity of time required for the annihilation of whole continents by so insensible a process."

Hutton's and Lyell's works were the origin of the idea of deep time, which continued to deepen as calculations got more precise. An 1890s estimate put the age of the earth at 100 million years. Two decades later, once the radioactivity in rocks could be used to date them, the earth was said to be 1.6 billion years old. Three billion seemed right in the 1930s, shortly after *Biosfera* was published. The now-agreed-upon age—based on dating earth's oldest rocks, and moon rocks brought back by astronauts, and samples from meteorites that fell to earth—is 4.54 billion years. Plus or minus 50 million. The biosphere is not that much younger, maybe 3.8 billion years old. Which means, environmental historian J. R. McNeill told *New Scientist*, that "we are all part of an unimaginably long chain of being, both human and non-human."

As the earth got older, life's lifeline also lengthened and deepened, and animals acquired a backstory of their own. In Hutton's day it was still assumed that every species was a permanent part of the landscape, a constant presence. Giant fossilized bones of creatures no one had ever seen, like mammoths and mastodons, had been unearthed, and there were theories about that: the bones might be what was left of the elephants Hannibal brought from Africa when he crossed the Alps to invade Rome. Thomas Jefferson thought it possible mastodons still roamed unexplored parts of America.

But in 1796, eight years after Hutton's boat ride, a twenty-six-year-old French zoologist, Georges Cuvier—who would later be proclaimed the founding father of paleontology (his is one of seventy-two names inscribed on the Eiffel Tower)—told the National Institute of Science and Arts that the massive teeth, recovered from mammoths and mastodons, were as different from modern elephants as "the dog differs from the jackal." Animals this size, he said, are nowhere to be found anymore; they're too big to be hiding themselves away; they no longer exist and will never return.

Absence, disappearance, discontinuity, extinction—all had to be accepted as fundamental components of life. Cuvier's view was that mammoths and mastodons (a name he coined a few years later; they were originally called the "Ohio animal") and giant ground sloths, whose bones were also dug up and reassembled, must have lived in "a world previous to ours, destroyed by some kind of catastrophe. But what was this primitive earth? What was this nature that was not subject to man's dominion?" Returning to this theme in middle age, Cuvier challenged his colleagues, reminding them that astronomers had "burst the limits of space," and he posed a new question: "Would it not also be glorious for man to burst the limits of time, and, by a few observations, to ascertain the history of this world, and the series of events that preceded the birth of the human race?"

A decade after Cuvier died, Sir Richard Owen, founder of the Natural History Museum in London, coined another name, "dinosaur," for a group of "terrible lizards" whose fossilized bones were being found in southern England. The world learned that, before mammoths or mastodons, reptiles, some of the largest land animals ever known, dominated the planet and then disappeared. Since then it's gotten easier to burst time's limits and think about the antiquity of life. From there it's not much of a stretch to catch up to Vernadsky, who postulated that "the biosphere has existed throughout all geological periods, from the most ancient indications."

Luca—as a given name it can mean "light" or "sacred wood." As an all-caps acronym, coined recently by geochemists, it stands for "last universal common ancestor"—a simple, fragile life-form, presumably a tiny protobacterium. LUCA carried 355 genes still found in the two earliest domains of single-celled life, bacteria and archaea. And since archaea gave rise to more complex cells with a nucleus, the basis for the proliferation of plants, animals, and all the rest of life, including slime molds and fungi, that makes LUCA the Eve of anything and everything alive.

It's thought that LUCA lived around 3.8 billion years ago, and since genes don't emerge overnight, life in some experimental form may have existed millions of years before that. Cracking open a grain-of-sand-sized zircon crystal in Western Australia, known to be 4.1 billion years old, geochemists found a "chemo-fossil," according to the Associated Press. It's a carbon trace that Mark Harrison, coauthor of the study, calls "the gooey remains of biotic life." Stephen Mojzsis, a geologist who describes his work as the search for the antiquity of the biosphere, told the AP that if this is a first indication of life on earth, "it arrives fast and early."

Claiming the biosphere

What could you call a place like the biosphere that only gets more extraordinary—more Vernadskyan—whenever you look at it more closely? A name that occurred to me is Conversation Hall, which I was introduced to by reading a 1982 piece in *The New York Times* by William Robbins: "For Hyman Myers, the experience at City Hall was like burrowing into an Egyptian tomb and uncovering a chamber of treasures. Hearing an excited shout from an assistant," Myers, a restoration architect, crawled through a dusty ceiling space in the massive, sumptuous City Hall in Philadelphia. After poking a flashlight into a black cavity, Myers punched through a thin plasterboard wall and uncovered Conversation Hall, an immense three-story-high room. As architecture critic Thomas Hine wrote in *The Philadelphia Inquirer,* it has "almost more detail than you can possibly take in" because there's "always more there than you expect": sandstone walls with marble accents in three colors, a green and lavishly gilded ceiling, busts of local greats by Alexander Calder's grandfather, a bright floral mosaic floor, and an enormous circular bronze-and-brass chandelier.

Inside the already gaudy City Hall building, which took thirty years to complete, Conversation Hall was hidden for almost that long by two inserted floors of drab, nondescript, 1950s cubicles, where people filled out forms with no idea of the splendors a few feet to each side, overhead, and underfoot. But it was intact, thanks to another heroic architect, George I. Lovatt Jr., who carefully protected everything when installing the cubicles. Now restored, the enormous room is available for large public meetings and for occasions like weddings.

What are the stumbling blocks that make it hard to see beyond the cubicles and claim the biosphere as our own? For one thing, there's its improbable shape, which as a geometric form is nameless and hard to visualize. It's not a slightly flattened near sphere (oblate spheroid) like the earth. The biosphere is the space *between* two spheres, and its boundaries aren't precisely mapped out. All we know is that the bottom of the biosphere is somewhere in the rocks below our feet (and beneath the oceans), while the top is somewhere in the air above our heads.

The only dimension we can measure with any exactness is its circular side-to-sideness, the seven continents and the five oceans, land and water—the surface area, 196.94 million square miles. It's been five hundred years since the Basque explorer Juan Sebastián Elcano completed the first circumnavigation of the globe, taking over after the death of Ferdinand Magellan ("You went around me first," proclaimed Elcano's coat of arms in Latin).

Somewhere along the line, as kids also get taught in grade school, if you travel in any direction long enough, east blurs into west, north into south. On a cloudless day, the curvature of the earth is only just barely visible from the windows of a commercial jetliner six miles up; for anyone on the International Space Station, 240 miles away from earth, it's a constant presence. From down on the surface, looking ahead, behind, left or right—the planet feels infinite, a finite unendingness.

Then look up—there's the thinness of the biosphere, the fact that, though three-dimensional ourselves, our entire lives are spent in something far closer to two dimensions.

The highest-flying birds, common cranes, migrate a mile over the Himalayas to avoid golden eagles cruising through the mountain passes below. Once, soaring even higher, was a Rüppell's griffon, a large mottled brown African vulture now considered critically endangered. It collided with a commercial jetliner seven miles above the Ivory Coast in 1973 (the pilot lost an engine but landed safely). Microbes have been found twenty-five miles

Rüppell's griffon, the highest-flying bird

straight up, in the above-freezing very thin air near the top of the stratosphere, and probably could live a bit higher. "Generally, people don't think of microbes being airborne," Shiladitya Das-Sarma, a microbiologist at the University of Maryland School of Medicine, told *Astrobiology Magazine.* "But there's a saying in microbiology: Everything is everywhere." If twenty-five miles is the top, and the bottom of the deep biosphere is at least seven miles down, then the total height of the biosphere is thirty-two miles, give or take a little.

Which means that all of life, all the species with two names that have been indexed, and all the uncounted, nameless, unnoticed ones out there, in addition to the processes that make life possible and that life contributes to and shapes, everyone and everything are sealed inside one age-old shelter, an impossible-sounding configuration that, at 32 miles tall and 24,901 miles around, is eight hundred times wider than it is high. From a distance it might look like a low huddled tent, but actually, as has been pointed out, it's not even thick enough to show up sideways from a space shuttle veering away from earth.

The preponderance of life exists only within the 12.5-mile-high attic-to-basement realm defined by the top of Everest (5.5 miles up) and the bottom of Challenger Deep, 7 miles down off the coast of Guam. This is a distance that John Gribbin, an English astrophysicist, has pointed out could be covered by car in less than twenty minutes on an open road. In the 1980s, Sir John Kendrew, a Nobel Prize–winning biochemist, called the biosphere the "thin green smear." Gribbin wrote in his book, *From Here to Infinity,* that if the earth were the size of a grapefruit, the green smear would be "a layer just 0.2 mm thick . . . no more than a coat of paint."

Within the thin green smear itself, there's the sea-surface microlayer, or SML. "The top five-hundredth inch of the ocean,"

Earthrise

says a *New York Times* report by Carl Zimmer in 2009, "is somewhat like a sheet of jelly . . . thinner than a human hair." It's a habitat of microbes all its own, visible in places. "Sailors," Zimmer says, "have long known that the surface can be covered with oily slicks (hence the phrase 'pouring oil on troubled waters')." This "fine coat of slime" is hard to retrieve and difficult to study—we've only just scratched the surface of it.

The most famous portraits of the earth from space—compelling images that once seen become indelibly fixed in the mind—generate intense feelings but have the disadvantage of showing the biosphere from the outside instead of edgewise, as something down there rather than what we're inside of. First came *Earthrise*, still shocking because it looks like a trick shot: in a black sky, a crescent earth, blue and white and glowing, moves up behind a much larger dull-gray moon. It's a picture taken out a window of a space capsule on Christmas Eve 1968. Writing in *The New York Times,* the poet Archibald MacLeish said that now people could "see the Earth as it truly is, small and blue and beautiful," a "bright loveliness in the eternal cold." For David Attenborough, his first sight of *Earthrise* was when "I suddenly realized how isolated and lonely we are here on earth."

Several years later, another astronaut on the way to the moon snapped *The Blue Marble,* the first clear image of the full round face of the earth bathed in sunlight. It shows clouds swirling over Antarctica, most of Africa, and parts of the Atlantic and Indian Oceans. It's been called one of the most widely reproduced photographs in history, and it's the original "whole earth" shot still printed on T-shirts and mouse pads. To Gregory Petsko, an American biochemist, the image was instantly iconic because, as he wrote in *Genome Biology,* "it perfectly represented the human condition of living on an island in the universe, with all the frailty an island ecosystem is prey to."

In 1990, Carl Sagan persuaded NASA to let *Voyager 1,* the first spacecraft to leave the solar system, turn around briefly and take the *Pale Blue Dot,* as Sagan called it, a farewell shot of earth as seen from almost 4 billion miles away, beyond Neptune. (Traveling outward ever since, *Voyager* is now 12 billion miles away.) Framed by a beam of scattered sunlight in an otherwise enveloping blackness, the earth is barely visible—"a one-pixel object," Ann Druyan, Sagan's wife and coauthor, puts it. "You can't look at that image and not think of how fragile, how fragile our world is." The dot, once you find it, is blue—earth's signature color.

These breathtaking images helped awaken modern environmentalism, a generation's unprecedented surge of interest in the health of the planet; curiously, there was no simultaneous surge of interest in the biosphere, the source of that health. The 1970s brought us the Environmental Protection Agency and the Endangered Species Act, and the 1990s saw international treaties, such as the United Nations Framework Convention on Climate Change and the Convention on Biological Diversity. But it's as though someone drew a Line of Demarcation through the oneness of it all, the way Pope Alexander VI in 1493 drew a line across South America to divide the continent into divergent Spanish and Portuguese realms.

In this case, the realms being divided were the environment and the biosphere. For many, the environment seems somewhere out there in a big world, something to separate from, set to one side. The biosphere is indissoluble, all-inclusive. It's our within-which, the one place we can never step out of or aside from (even astronauts take it with them). Ultimately it's the only "here" anyone has. And since in cosmic terms it's no thicker than a green smear on a microscope slide, there's little room for error.

Gaia

In a 1785 Edinburgh lecture, James Hutton, several years before his deep-time boat ride, had another *Biosfera*-like idea: "I consider the Earth to be a super-organism"—a word he coined on the spot—"and that its proper study should be by physiology." As if it had the health and equilibrium of a living being, along with component parts like organs, muscles, and arteries. The idea took deep time (two hundred years) to get noticed by Western science. In the 1970s, after *Earthrise* and *The Blue Marble*, James Lovelock, a British inventor and futurist with a PhD in medicine, wrote about a vision of the planet as a complex superorganism that takes care of itself, and us.

Lovelock called his theory "biocybernetic universal system

tendency/homeostasis"—"homeostasis" referring to an organism's ability to keep itself stable. William Golding, the Nobel Prize–winning author of *Lord of the Flies*, and Lovelock's neighbor, persuaded him to rename it the Gaia hypothesis, after the Greek goddess of earth.

In a 1995 essay, evolutionary biologist Lynn Margulis, who worked on Gaia with Lovelock, said, "Gaia is a tough bitch— a system that has worked for over three billion years without people." But Margulis herself preferred to say that the earth is, in her words, "an ecosystem, one continuous enormous ecosystem." Margulis said Lovelock thought the word "organism" would resonate: "Let the people believe that Earth is an organism, because if they think it is just a pile of rocks they kick it, ignore it, and mistreat it. If they think Earth is an organism, they'll tend to treat it with respect."

Gaia took on life, the idea spreading widely and quickly. It was the inspiration for the video game *SimEarth;* in Isaac Asimov's science fiction novels *Foundation's Edge* and *Foundation and Earth*, Gaia is a planet that aspires to become Galaxia, a superorganism spanning the Milky Way. Lovelock's books continue to be best sellers. But the scientific community viewed Gaia skeptically—"Our planet is less robustly stabilized than Gaia implies, and therefore more fragile," was the verdict of oceanographer Toby Tyrrell in *New Scientist*. This, even while seeing air, water, land, and life as an integrated system. Today, with no William Golding next door to help out, these studies are unpoetically called "earth system science," or ESS.

The insights of chemists, physicists, biologists, and mathematicians are all integral to ESS, as are those of ecologists, economists, geologists, glaciologists, meteorologists, oceanographers, paleontologists, sociologists, and space scientists. NASA set up an ESS Committee in the 1980s. "Studies of the continents, oceans, atmosphere, biosphere, and ice cover over the past thirty years," NASA wrote, "have revealed that these are components of a far more dynamic and complex world than could have been imagined only a few generations ago." It took forty years, but in 2010, the director for planetary science at NASA, James Green, acknowledged a debt to Gaia, telling a conference of exobiologists, "Dr. Lovelock and Dr. Margulis played a key role in the origins of what we now know as earth system science."

Every 26 million years

As time deepened, there were those who, like Hutton and Lyell, believed that changes to the earth, such as erosion and the forming of mountains, came slowly, continuously, visibly, and predictably—"The present is the key to the past," as Lyell put it. These were the gradualists, also known as uniformitarianists or incrementalists. They were challenged by the catastrophists, like Cuvier, who said fossils of extinct creatures could only be explained by sudden cataclysms that quickly annihilated "former worlds." Having just lived through the French Revolution, Cuvier thought of these convulsions as earth's own "revolutions."

Still, there appeared to be no evidence, in the ground or elsewhere, for any such upheavals, and Darwin's view was that the evolution of life was a gradually unfolding process. Catastrophism was considered unscientific, an outmoded belief. In the words of Angelo Heilprin, a nineteenth-century geologist and mountaineer (a glacier in Greenland is named for him), "It is illogical, and directly opposed to the workings of evolutionary force, to conceive of a wide-spread group of animals suddenly appearing and springing into prominence; and no less illogical to conceive of an equally sudden extermination."

It was known, however, that dinosaurs disappeared suddenly

66 million years ago—but why? The theories put forward were thin. Flowering plants appeared sometime in the 180 million years dinosaurs were dominant, so maybe dinosaurs succumbed to hay fever, or developed cataracts during lifetimes out in the sun, causing them to fall off cliffs.

But in 1980, Luis Alvarez, a Nobel Prize–winning physicist, his son, Walter, a geologist, and chemists they worked with showed that rocks all around the world held evidence that 66 million years ago, a comet or asteroid six miles wide and with a speed of 45,000 miles an hour (about twenty-five times as fast as a bullet) struck the earth, setting off tsunamis and throwing enough dust in the air to darken the skies and shut down photosynthesis. The impact would have had a force 2 million times greater than Tsar Bomba (Russian for "the King of Bombs," the most powerful hydrogen bomb ever test-fired, three thousand times more explosive than the atomic bomb dropped on Hiroshima). At the impact site itself, the Alvarezes said, there must be a crater at least one hundred miles wide.

The Alvarez hypothesis had its own explosive impact. The Alvarezes' June 1980 paper in *Science* magazine, says biologist Sean B. Carroll in his book, *Remarkable Creatures,* is "perhaps unmatched in scope by any other single paper in the modern scientific literature." For Egyptian paleoecologist Ashraf M. T. Elewa, who edited and contributed to the book *Mass Extinction,* it was the Alvarezes who presented the "stunning and convincing mechanism" that finally explained the dinosaurs' extinction.

On the other hand, a sticking point: the Alvarezes did not produce a crater of the right size and age. Then, in 1991, three years after Luis Alvarez died at seventy-seven, other researchers introduced the world to the Chicxulub crater, buried twelve miles below a Yucatán village of that name. Chicxulub means "the devil's tail" in Yucatec Maya. The crater had been known about since 1978, even before the Alvarezes published their findings, but it took a dozen years before anyone realized that it matched almost exactly what the Alvarezes had described: a crater formed by an asteroid or comet six to nine miles wide that hit the earth just less than 66 million years ago. Chicxulub crater is 111 miles across.

Terms and catchphrases grandly coined by geologists who couldn't yet date things exactly spoke of the Age of Reptiles giving way to the Age of Mammals, and the Mesozoic Era being supplanted by the Cenozoic Era (when "middle life" no longer held sway and "recent life" flourished). All now stood in vivid relief. Three-quarters of the species on earth disappeared along with the dinosaurs. It was a mass extinction. There was now a firm end date for one of Cuvier's "former worlds."

In an extraordinary burst of research, further lost worlds were recovered and dated just in the interval between the Alvarez hypothesis and Chicxulub. This acceleration came about thanks to Dr. J. John—he preferred Jack—Sepkoski Jr., a paleontologist and something of a math whiz. Sepkoski and colleagues at the University of Chicago (for some reason nicknamed the Chicago Mafia) provided the missing evidence.

"Ten Years in the Library," Sepkoski called one of his essays, citing library shelves as his "field site." For generations, paleontologists, whom other biologists derided as stamp collectors, amassed thousands of distinct monographs about fossilized marine invertebrate animals that were no more, with birth and death dates—their earliest and most recent appearances. As an explorer of the "natural history of data," Sepkoski amalgamated this scattered information, recording these dates into a computer (something no one else had done, or perhaps had the patience for), all while listening to Spinal Tap and the Sex Pistols. He created the world's first database of extinction—more than thirty thousand entries in all.

Modern paleontologists—computer jockeys and number crunchers—are "about as different from Indiana Jones as you can get," says Sepkoski's son, David, a science historian, in an essay published in *Aeon* magazine. But, he adds, there's drama enough in the database: "to expose patterns in the history of life that emerged only on very long timescales," leading to the discovery that life has "experienced major, catastrophic mass extinctions at least five times in the Earth's history (this is why many people now refer to the current biodiversity as the 'sixth extinction')." It's estimated that it takes the planet 15 to 30 million years to recover from such an event. Elizabeth Kolbert used *The Sixth Extinction* as the title of her Pulitzer Prize–winning book.

Jack Sepkoski and a colleague, David Raup, wrote a paper ("Periodicity of Extinctions in the Geologic Past") that also identified twelve extinction events over the past 250 million years, which for unknown reasons showed up with what Sepkoski called "clock-like behavior," one after another, every 26 million years. Sepkoski's obituary in *Nature* magazine calls this finding a bombshell that "created waves well beyond the paleontological community." It's still hotly disputed, hotly defended. Jack Sepkoski died in 1999, at age fifty. "Most of us aren't going to be remembered in a hundred years' time," paleobiologist Douglas Erwin told *The New York Times,* "but Jack will, because he changed the way we think about the fossil record and the history of life."

David Sepkoski grew up in the 1980s, the last decade of the Cold War, and remembers the fear of nuclear war, where there would be no victory or defeat but instead a long, cold, sky-darkened "nuclear winter." He says in a 2015 book, *Endangerment, Biodiversity and Culture,* "If the dinosaurs could go, the idea went, then so could we humans." And not only humans—a "new catastrophism." In short, Cuvier brought up-to-date: all of everything under siege, animal habitats and spectacular landscapes.

The first National Forum on BioDiversity—the word, coined in 1985, then had two capital letters—was a four-day event in Washington, D.C., in 1986, organized by E. O. Wilson and Walter Rosen, a botanist. Raup, a main speaker, showed that the "normal" or "background" rate of extinction is two or three species a year. Here, now, was a baseline for comparison. A "contemporary extinction problem," Raup said, emerged from the database in a clear, inescapable way—and it would be as irresponsible to ignore it as if "an epidemiologist were to treat an infectious disease without medical records." For the hundreds of people at the event, the big takeaway (written up in a group statement) was that "the species extinction crisis is a threat to civilization second only to the threat of thermonuclear war."

The second story

Today the global extinction rate is a thousand times higher than the normal or background rate, and the United Nations Environment Programme estimates that 150 to 200 species go extinct every day. Which means the year the Sorting Hat spider and 18,000 additional species got discovered may have seen the loss of 73,000 others. The "within-which" that holds all of life, the green smear, isn't any thinner, but the "among-which," the abundance of species and their ability to move around, is shrinking. (Technical terms for these are defaunation and loss of vagility.) We're living through what Vernadsky anticipated long before "BioDiversity" became a concern. He wrote (as translated in a 150th birthday tribute by Russian geochemist Mikhail Marov):

> Mankind, taken as a whole, is becoming a powerful geological force. . . . For the first time, man's life and his cul-

ture encompass the entire upper envelope of the planet—in general, the entire biosphere. . . . Man has actually comprehended for the first time that he is an inhabitant of the planet and may—must—think and act with a new perspective, not solely with the perspective of a single individual, family, or clan, or of nations or alliances among them, but with a planetary perspective.

Vaclav Smil stresses this can't happen soon enough, since, he says in *The Earth's Biosphere,* "few challenges facing civilization during the twenty-first century will be as daunting, and as critical, as the preservation of the biosphere's integrity." Will a biospherical perspective ever become . . . second nature? Smil says Vernadsky himself was hopeful, writing in the last year of his life, "I look forward with great optimism. I think that we are experiencing not only an historical change, but a planetary one as well. We live in a transition."

It's been five hundred years since the Polish astronomer Nicolaus Copernicus demonstrated that the heavens don't revolve around the earth, and five hundred years since Magellan and Elcano sailed the globe. But according to a 2018 YouGov online poll, 2 percent of Americans still believe the world is flat, and a National Science Foundation survey conducted in 2012 found that 26 percent believe the sun circles the earth. Since we live inside a layer of life too thin to be seen from space—the cosmic equivalent of the five-hundredth-of-an-inch-thick sea-surface microlayer—how can we get a sense of it?

There's a place to see the concept of deep time up close. A cast of a rock is on display at NASA's Goddard Space Flight Center, just beyond the Washington Beltway. It's a re-creation of a 110-million-year-old red sandstone slab that was almost pulverized by jackhammers when the center was undergoing expansion.

It's eight and a half feet long, covered with seventy muddy tracks all made within a week, or maybe two, or maybe over a few hours. A young sauropod, a plant eater with a long neck; a baby nodosaur, an armored plant eater, following its parent; flying pterosaurs walking around, searching for food; and mammals the size of squirrels being stalked by relatives of the *T. rex* no bigger than ravens.

All this produced "ecstasy" in Ray Stanford, the amateur paleontologist who found the rock in 2012. He told *The New York Times,* "To see them with their potential predators"—dinosaurs scoping out mammals to eat, something paleontologists had never seen before—"that hit me for a loop."

To summon the sense of the within-which, to leave the mental cubicles and step into Conversation Hall, is to remember that wherever you happen to be, you're poised between the rock bottom and the uppermost layer of the stratosphere—and the floor of life is closer than the ceiling. Humanity's place, the caribou's, along with the tiger's and the sequoia's, is on the second floor of this grand, squat, unique four-story building—unless you're in a jet, in which case you're briefly on the third floor. We're at the top of the food chain, but we're also, like so many other species, second-floor citizens. The setting for humanity's story is just that—the second story.

The Science of 50 Percent

Why half? Why zero in on saving precisely 50 percent of the planet's land (and water too for that matter)? More essentially, "How much is enough?" Reed Noss has been asking this question for more than thirty years in a series of books and essays; he's a conservation biologist who thinks at a continental scale. In his work, and in that of many other conservation biologists, there's no one percentage that holds true everywhere and for everything, but 50 is a good stand-in for perfect. It's at the midpoint of a range of targets put forward by dozens of scientific studies, with low estimates near 25 percent and high ones exceeding 75 percent.

Back in the early twentieth century, pioneering but neglected studies produced similar results, though they made so little impression they weren't even shouted down. Imagine if they had been listened to. The extinction problem might have been solved even before it emerged—50 percent by 1950. But that alternate history never had a chance, Noss says. The world wasn't ready to start thinking bigger.

Take long-neglected Warren H. Manning, described by architectural historian Christine G. O'Malley in *Landscape Architecture Magazine* as having been "puzzlingly understudied for years." What's generally known about Manning is that he was a landscape architect trained by Frederick Law Olmsted, the father of landscape architecture; Manning himself was a founder of the American Society of Landscape Architects and later its president. He was sensitive to his surroundings and in his work expressed the essence of a landscape by using only native plants—he called his projects "wild gardens."

With his talent and background, Manning became so successful he designed more than 1,700 sites, taking on work of all kinds: palatial estates for the founder of Goodyear Tire and Rubber Company and for beer barons like the Pabst and Busch families; grounds of small homes; suburban subdivisions; industrial towns; public parks; college campuses; cemeteries; golf courses. He was in demand in thirty-six states at the historic crossover moment when the growing U.S. population was becoming more urban than rural.

In 1917 new contracts slowed as the country entered World War I, and Manning, with time on his hands, saw this as an opportunity to make a postwar difference. His idea, he noted, was to "make prosperity dominant and to minimize adversity" when peacetime growth resumed by eliminating waste of the nation's resources: "soil, forests, oil, gas, coal, minerals and wildlife." Manning wrote, "If we want our country to stand first among all nations then very many more of our citizens must gain a better knowledge of this United States as a whole."

With posterity as his only client, Manning enlisted thirty workers in his Massachusetts office to help draft a National Plan, 427 pages long with 320 maps, graphs, and drawings. It's a marvel of foresight and vision in which he basically treated the whole country as if it were a single landscape he was commissioned to design a future for.

Borrowing an Olmstedian technique, landscape historian Robin Karson has noted, Manning and his team created "overlay" maps of soils and forests, topography, and other features— "fur, fin, and feathered creatures, as well as the hopping and crawling creatures without fins, fur, or feathers," he rather airily called them. When amalgamated, by being superimposed on one another, the maps make clear where to build and where to leave things open. Now computerized, such "stacked maps" are a standard part of modern geographical information systems (GIS). But back then Manning and company had to generate every image by hand, delineating nine kinds of soil and five animal and plant zones, all overlapping state lines on both coasts and in the Great Mississippi Basin in between.

At the time, there was only one transcontinental highway in the country, the two-lane Lincoln Highway from Times Square in Manhattan to Lincoln Park in San Francisco—driving on it, according to the 1916 *Complete Official Road Guide of the Lincoln Highway,* was "something of a sporting proposition." But forty years before the Interstate Highway Act, Manning anticipated the need for a network of coast-to-coast and intersecting Canada-to-Mexico "great trunk-line thoroughfares." As part of his "wild garden" approach, he says in an unpublished autobiography, he wanted to see "special floral, fruit, or autumn color plantation" along these roadways. But he got part of the future wrong, expecting cars to be going no more than thirty miles an hour: "One would cover a mile in about two minutes, and this hardly gives time to enjoy or appreciate such beauty."

The National Plan was put together only fourteen years after the Wright Brothers successfully got their spruce-wood biplane off the ground for the first time at Kitty Hawk (the longest flight that day lasted fifty-nine seconds), and Manning assumed that flying, like driving, would be part of the everyday commute. He wrote: "There will be airplane landing strips at intervals along the sides of these straight thoroughfares where people can pass from the plane to the automobiles . . . or to the autogiro taxicab type of aircraft that will taxi people from the main field landing places to their destinations."

The seed of Half Earth in the National Plan was Manning's insistence that "one third of this continent is to retain its natural aspect" as a "National Re-Creation System" of "inevitable wildernesses." Public reserves would be linked by trails and scenic "recreation ways." Land would also be set aside as stopovers for migrating birds—"our original aeroplanes."

Except for this concern for birds, Manning's pro-wilderness arguments are very much pre-biodiversity, centering on what nature does for people. Essentially he was carrying forward what Olmsted said back in 1865, about why Yosemite Valley needed to be preserved: "It is a scientific fact that the occasional contemplation of natural scenes of an impressive character . . . is favorable to the health and vigor of men and especially to the health and vigor of their intellect." As Manning put it, *We shall some day evolve a definite program for preventive, rather than local curative measures, for most of our ills physical, mental and moral, and the great out-of-doors will be the great builder and re-creator.* The italics are his.

The National Plan found one powerful admirer, Franklin K. Lane, Woodrow Wilson's secretary of the interior, who supported the creation of the National Park Service despite feeling nothing for nature. "A wilderness," Lane said, "no matter how impressive and beautiful, does not satisfy this soul of mine, (if I have that kind of thing). It is a challenge to man. It says, 'Master me! Put me to use! Make me something more than I am.'" But Lane wrote about the plan, "The making of America is the task that

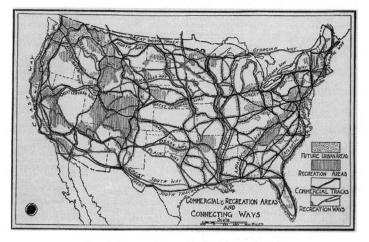

From Warren Manning's National Plan

has challenged the vision and the energies of Americans for 250 years. We have gone at this work haphazard and done a thing that is marvelous. But now has come the day when we must see the problem as a whole and plan to meet it scientifically. This Mr. Manning urges, and he has made a study of what America is that is fascinating as fiction, yet it is all very solemn fact."

Except when Lane showed the National Plan to his staff, it was energetically rejected. One geologist said if it showed changes to state lines (which it did), "he would throw it away and not read it." *Landscape Architecture Magazine* printed a twenty-four-page extract in 1923. Manning died in 1938. A 2017 book celebrates some of Manning's surviving landscapes. The National Plan itself remains unpublished.

Buffers

Warren Manning was said to have spent his working life continually in transit, always writing and sketching, typically on a train.

I like to think he might well have crossed paths with a younger man, Victor Ernest Shelford, the father of animal ecology in North America. After a 1910 trip to the Rockies and California, Shelford's quest was to explore every major habitat on the continent, as George O. Batzli and Jeffrey D. Brawn explain in their chapter on Shelford in *The University of Illinois: Engine of Innovation.* Shelford did so, establishing himself as a force of nature in the emerging field of ecology. Photos show him with a penetrating if slightly quizzical glance. Shelford insisted on punctuality and discipline; people spoke of him as "highly organized, industrious, dedicated, tenacious, and fearless." Some added "aggressive." He would tell people exactly what he thought of them, whatever their rank.

Shelford studied lemmings, and snowy owls, and chinch bugs (a pest that eats lawns), and codling moths, whose caterpillars are the "worms" that tunnel through apples. He spent ten years perfecting underwater photoelectric cells to examine how light penetrated seawater, to get a better idea of how deep different algae would grow. And long before global warming became a concern, he started a "century-cycle" in Trelease Woods at the University of Illinois, to measure animal populations year after year, for one hundred years, to see how they're affected by changing weather patterns.

For decades Shelford led expeditions with a slew of grad students, who remembered these transcontinental field trips— heading north to Hudson Bay, or south to Panama—as arduous but the making of them. Shelford's first book, *Animal Communities in Temperate America,* written in 1913, is one of the first great ecological studies. His last book, *The Ecology of North America,* came out in 1963 when he was eighty-six and was praised by *Science* magazine as "the impressive climax to a half century" of passionate research. When Shelford died in 1968, at ninety-one, John D. Buffington, an entomologist (insect specialist), wrote,

"Not only his students, but also his students' students are among the most distinguished of today's ecologists."

Born on a farm in upstate New York, Shelford didn't graduate from high school until he was almost twenty-two—but then he roared ahead, except, he admitted, for never learning how to speak well in public; critics said his lectures lacked polish. His 1913 book transformed him from a zoologist into an ecologist. He had been studying tiger beetles, fish, and other animals, showing how the success or failure of any animal could only be understood by connecting it to its surroundings, in this case the sand dunes on the shores of Lake Michigan. (Tiger beetles, the size of a pinky finger, are said to be as fierce and fast as a cross between a tiger and a cheetah.)

From this, Shelford came up with his law of tolerance, a kind of Goldilocks principle—to thrive, tiger beetles and other organisms had to withstand the worst of things, too much heat and freezing temperatures, while also not drowning or getting dried out. By 1915, Shelford helped organize the Ecological Society of America (ESA) and was elected its first president.

By 1921 Shelford, as head of the ESA Committee on the Preservation of Natural Conditions, had identified six hundred areas worth protecting, and in 1933 he published an article that was his own National Plan for nature, a network of "Nature Sanctuaries or Nature Reserves." Like Manning's insistence on the permanence of "inevitable wildernesses," Shelford's idea of "a nature sanctuary with its original wild animals for each biotic formation" is now thought of as one of the great if-onlys of pre–Half Earth days: conservation biologist R. Edward Grumbine, author of *Thinking Like a Mountain,* wrote, "One can only dream of what condition US public lands would be in today if policymakers of the time had embraced Shelford's bold vision. But no sanctuary system was forthcoming."

Although Shelford's goal was to protect animals, plants, and entire ecosystems, he (again like Manning) felt that human need was the main reason for it. "Preservation of Natural Conditions," a 1922 article by Shelford and his ESA committee, quoted the provincial botanist of British Columbia, John Davidson, affectionately known to Canadians as Botany John: "Just as we preserve the works of great masters, and find that the longer we have preserved them, the greater their value becomes; so we are seeking to preserve the works of the greatest of Masters, and if length of time increases the value of these works they are infinitely more valuable than works of art. In this we appeal to almost all sections of the community."

Davidson then ticked off the very practical reasons for conservation. Those in business, he said, liked the money spent by visitors to wild places. Artists and poets considered them a "source of inspiration," teachers a "source of illustration," students a "source of instruction." And for all, said Botany John, they were "a source of health and recreation which leads one's thoughts away from the mundane affairs of this world 'Through Nature up to Nature's God.'" To which Shelford added: "It pays to preserve forests and swamps as watershed protectors and flood preventers."

No takers—and yet. Maybe we can think of the Shelford plan, ignored as thoroughly as Manning's, as its own kind of century-cycle, preparing the ground, if you like, for 50 by '50. There's a profoundly original suggestion in his sanctuary plan that's so exact it sounds like something out of an instruction manual. Shelford wrote, in a 1933 note in *Science* magazine, "The nature sanctuaries are surrounded by areas in a less natural state, called buffer areas of partial protection. In a buffer area the vegetation is only slightly modified by man. It is a region of partial protection of nature and is zoned to afford suitable range for roaming animals under full protection."

This was Shelford's way of trying to respect human *and* animal needs, a concept at the very core of Half Earth. Shelford assumed the logical place for animal sanctuaries would be inside national parks or forests—public land that would now be closed to people. But the larger "roaming animals" (he singled out wolves, mountain lions, bobcats, and coyotes) wouldn't understand the rules, and from time to time, he anticipated, they would wander beyond the sanctuaries, and maybe outside the park altogether.

A buffer on private land would make the boundaries less rigid and give the animals that much more room. It's not exactly the peaceable kingdom foretold by the prophet Isaiah 2,700 years ago ("And the wolf will dwell with the lamb, and the leopard will lie down with the kid"), but it might act like a demilitarized zone, an area of equal footing. Forty years later, and seemingly overnight, skipping even a beta stage, buffer areas became the great innovation for a new global conservation strategy, the World Network of Biosphere Reserves (WNBR).

Today, the 701 biosphere reserves in 124 countries ("special places," UNESCO, their sponsor, calls them) are areas where social and ecological systems overlap. Their motto: "Three zones, one biosphere reserve!" A meticulously protected core (first zone) is surrounded by a buffer (second zone) with a double task—conservation and accommodating a steady presence of local residents. Outside the buffer is what UNESCO calls a "transition area" (third zone), still within the reserve, with larger populations and more development, where "the greatest activity is allowed."

A Shelford-like idea from Reed Noss, first set down in a 1987 essay, "Protecting Natural Areas in Fragmented Landscapes," offers another dimension, pointing out a way to think beyond individual reserves and across a larger landscape. The buffers between the wildest places and the rest of the world don't have to be fixed and static. You can leave the core areas in place but reshape, stretch, and lengthen the buffers so they don't just sit

Reed Noss at Myakka River State Park, Florida

there enveloping the cores. They can take on new purpose, like tendrils seeking sunlight, heading out across the landscape for hundreds of miles, along riverbanks and through lightly used backwoods, until meeting up with other reserves.

That way, Noss thought, you could for instance link up 630,000 acres on the Florida-Georgia state line (Osceola National Forest and the adjacent Okefenokee National Wildlife Refuge, America's largest blackwater swamp, where the water looks like steeping tea) with the 1.5 million acres of Everglades National Park at the southern tip of Florida. This would turn the whole of Florida into a network of sanctuaries, and allow large roaming animals, like the Florida black bear and the Florida panther, to move over long distances; in 1985 there were only isolated pockets of bears and a couple of dozen panthers left.

Noss and his PhD advisor, Larry D. Harris, author of *The Frag-*

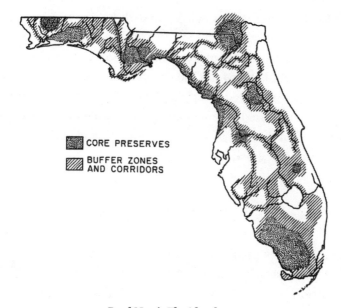

Reed Noss's Florida plan

mented Forest, set about designing a network of wildlife corridors weaving throughout Florida—maybe without anyone noticing, because these corridors would parallel the built-up parts rather than interrupt them. This meant doing what Manning did seventy years earlier, creating "stacked maps," though this time with plastic sheets instead of tracing paper, and better pens. "I made a pretty simple map of Florida," Noss told Mary Ellen Hannibal, who wrote about linking up landscapes in the Rockies in her book *The Spine of the Continent: The Race to Save America's Last, Best Wilderness.* "GIS was in use by the early '80s but I didn't know anyone using it; we used Mylar overlays and Magic Markers."

Noss's and Harris's drawings, which proposed protecting about 60 percent of Florida, got an unusual amount of attention—at the meeting where Noss and Harris made their ideas public, a candidate for governor happened to turn up. "My map appeared everywhere," Noss told Hannibal. "At first there was some backlash: this is some crazy, radical idea. But within a couple of years the state agencies in charge of acquiring land for conservation started using and refining my map." So far, about 10.5 million acres in the network are protected, nearly a third of the state. Panther numbers have increased ninefold, and a group called Florida Wildlife Corridor has the goal of accelerating the rate of conservation by 10 percent every year.

Nature for its own sake

Shelford shows up, if only in spirit, in a 1972 essay Noss points to as "the first publication that explicitly offered a science-based protection target" of 50 percent, which makes the science of Half Earth half a century old. It's a paper by the Odum brothers— Eugene P., one of the students Shelford took on field trips in the 1930s, and Howard T., eleven years younger. To colleagues they were Gene and Tom. The Odums have been called the two most important ecologists of the twentieth century, and Gene, with Tom's help, wrote the first ecology textbook, which has been translated into twelve languages.

As "ecosystem ecologists," they helped ecology become what's been described in the book *Ecosystem Ecology: A New Synthesis* as "a much broader church." They looked at how the physical and chemical, nonliving (abiotic) parts of ecosystems—nutrients like nitrogen, for instance—cycled through animals and plants alike.

The brothers were "fond of quoting and using one another's work in an almost symbiotic manner," Mark Glen Madison, a science historian, has written. Yet they were very different people, with Gene known as "your typical southern gentleman" (remembers Howard S. Neufeld, an ecologist), while Tom was brash, outspoken. Gene's strength, according to his brother, was "feeding ideas to people at a rate they could take them, and not shocking them. Whereas my tendency is to say what is correct, even putting down stupid ideas."

Their essay, "Natural Areas as Necessary Components of Man's Total Environment," is both sober and frank, and, Noss thinks, overlooked. The Odums take a Shelford idea—that forests and swamps are watershed protectors and flood preventers—and turn it into the strongest possible statement of why every area not yet built on has got to be part of a buffer and must be protected, for humanity's sake. "Our theme is that the natural environment is an essential part of man's total environment. Preservation of a substantial portion of the biosphere in a natural state, while not a panacea for all the ills of mankind, is nevertheless a necessity." Not to mention the advantages to towering cities, no less a part of the biosphere: "Specifically, cities need the protection of an adequate life-support system, many elements of which natural environment provides free of charge. Without natural recycling and other works of nature, the cost of maintaining quality life in cities would be prohibitive."

The Odum brothers chose South Florida as a case study and analyzed the flow of energy and materials between what Gene later called the "two houses of man"—the "house of nature" and the "man-made house" (cities). Ecosystems provide benefits (now called "ecosystem services" and "nature's contributions to people") that are irreplaceable—no amount of money or skill can duplicate them. Gene said in a speech on April 22, 1980 (the tenth anniversary of Earth Day), "We need now to give special attention to the life support buffer provided by natural and semi-natural ecosystems. It is the forests, prairies, lakes, rivers, estuaries, and oceans that make it possible for cities and industries to function." Unless the buffer is big enough, both houses will collapse.

In 1997, New York City launched the world's most famous ecoproject when it kept its drinking water pollution-free by allowing the environment to do all the work. The city had studies showing that bacteria can only travel 150 feet through soil. To be on the safe side, the city spent $438 million (as of 2015) to protect more than 135,000 acres of land to buffer upstate reservoirs and to keep development away. Nine million people drink this water, safeguarded by a watershed forest that's nearly 40 percent protected. The alternative? Spend $10 billion or more to build a filtration plant—same result, plus at least another $100 million a year to maintain.

The Odum brothers' model showed there needed to be "a 1:1 ratio of natural to developed environment" and they made a Half Earth recommendation: "It would be prudent for planners everywhere to strive to preserve 50% of the total environment as natural environment." Gene followed his own prescription, in life and in death. He was a professor at the University of Georgia for forty-four years, and after he passed away, twenty-six acres he owned in Athens, Georgia, were subdivided and developed, with 57 percent of the property set aside as open space.

In 1945, the Ecological Society of America decided it would be inappropriate for it to "influence legislation." Which led Shelford to form a breakaway group, in 1946: the Ecologists' Union. By 1950 it had a new name, The Nature Conservancy. With a million members, it has become this country's largest environmental nonprofit and protects more than 119 million acres around the world. Its mission: "A world where the diversity of life thrives, and people act to conserve nature for its own sake and its ability to fulfill our needs and enrich our lives."

This was new—nature for its own sake. Early Half Earth champions, though passionate about the land, always put people first. They talked about people's needs, as well as people's comfort, delight, health, safety, and survival; the perspective is that of a planet that has human beings as its primary purpose. They diminished the role of Gene's house of nature, reducing it to something like a back porch tacked onto the man-made house.

But there's a view from outside, one looking back at the lighted windows of the man-made house, a view that focuses on animals and plants and on protecting biodiversity for the sake of all life. This is a vision that's appeared far more recently, not focusing on how much space nature needs to keep people going but instead on how much land and water the millions of other species need so as not to get crowded off the planet. It's a vision that began with islands and the species found there.

Patterns

There seems to be no pattern to how and when a pattern can get noticed in a big way. "Anyone familiar with the history of science," said Robert H. MacArthur, coauthor (with E. O. Wilson) of *The Theory of Island Biogeography,* "knows that it is done in the most astonishing ways by the most improbable people and that its only real rules are honesty and the validity of logic."

"Without bold, regular patterns in nature," the British ecologist Sir John Lawton wrote in 1996, "ecologists do not have anything very interesting to explain." MacArthur had the same thought: "To do science is to search for repeated patterns, not simply to accumulate facts." MacArthur also told Wilson that he would "rather save an endangered habitat than create an important scientific theory," but their work together brilliantly accomplished both. It explored the oldest ecological pattern ever discovered, one showing how many species you could expect to find on any island. MacArthur was thirty-seven when the book was published and died only five years later, in 1972. A memorial tribute said he indelibly affected the basic beliefs and frontiers of ecology, frontiers that stretched back centuries.

"Dogmatic, humorless, suspicious, pretentious, contentious, censorious, demanding, rheumatic, he was a problem from any angle" is how New Zealand historian J. C. Beaglehole summed up Johann Reinhold Forster, an ordained minister who spoke or read seventeen languages and was the naturalist on Captain James Cook's second voyage around the world, which left England in 1772. Forster was eclipsed by his own seventeen-year-old son, Georg, who accompanied him on that voyage. At twenty-two Georg published an account of the trip that brought him fame and praise as a superb essayist and the founder of modern travel literature; the young Forster's book influenced Goethe and Alexander von Humboldt. The senior Forster constantly complained about his cabin and got mercilessly mimicked by the crew, and Cook later exclaimed, "Curse the scientists, and all science into the bargain!" Nonetheless, Forster became a respected professor at a German university and is known today as the patriarch of geography.

On Cook's second voyage, Johann Forster noticed that as the ship made its way through Polynesia, thousands of islands in the South Pacific between New Zealand and Hawaii, the variety of plants, not just the number, increased when the islands got bigger—as he put it, "Islands only produce a greater or lesser number of species as their circumference is more or less extensive."

This is the first recorded instance of what is now called the "species-area relationship." In the mid-nineteenth century, it was confirmed by, among others, a Forster-like British botanist, Hewett Watson, praised by Darwin but remembered by a contemporary, R. D. Meikle, as a "turbulent figure, a born controversialist, a pungent critic, and a most enthusiastic disturber of the peace" (i.e., a royal pain). Watson noted that a single square mile of Surrey, the most species-rich English county, had just about half the species in England as a whole. He then studied the plants in the rest of southern England, and finally in all of Great Britain. Increases popped up reliably—bigger areas, more species.

"Darlington's rule of thumb," as it came to be called, is a small chart in a 1957 book by Philip Darlington, a Harvard zoologist. He took Forster's and Watson's rough calculations into the realm of theoretical ecology, territory pioneered by the Odum brothers, using observations to lead to precise predictions.

In a way, Darlington was Forster's opposite—modest in his life but formidable in his work. His field trips around the world added thousands of specimens to Harvard's Museum of Comparative Zoology. His particular passion was carabid beetles, with at least forty thousand known species. In New Guinea, a giant crocodile clamped him in its jaws and dragged him to the bottom of a jungle pool. "Those few seconds seemed hours," he said. The crocodile, for some reason, released him, and Darlington scrambled to the shore and hiked to a hospital, weak from blood loss. He wrote home to his wife that he'd had "an episode with a crocodile"—and left it at that.

In the Caribbean, Darlington studied amphibians and reptiles from the biggest island, Cuba, to one of the smallest, Saba, no more than the top of a dormant volcano. From this he created his rule—namely, that if you survey an area, and then survey another one that's ten times bigger, you'll find twice as many species. Saba had five amphibian and five reptile species; Montserrat, about ten times its size, had nine of each, including the giant ditch frog, known as the mountain chicken; Cuba, about ten thousand times as big as Saba, had 76 amphibian and 84 reptile species. The rule would have predicted 80 of each.

MacArthur and Wilson endorsed Darlington's math in their

1967 book, *The Theory of Island Biogeography,* and showed it to be part of a larger pattern, this one cyclical and ancient. Here was an origin story for the species-area relationship, about colonization and displacement, accommodation and equilibrium, an Olympian view of epic, ever-repeating struggles. "The MacArthur-Wilson theory was a big hit," said Fred Powledge, a science historian, in *BioScience* magazine. "The term 'new paradigm' was heard throughout the hills and valleys of ecological research." MacArthur referred to himself as a mathematical naturalist, and at the heart of the theory is an elegant mathematical model of entrances and exits that he devised; its pared-down structure echoes what the philosopher Karl Popper called "the austerely beautiful *simplicity of the world* as revealed in the laws of physics."

Its essence can be seen in a minimalist graph (the x axis is time and the y axis is number of species) with only two curves. For Wilson, looking back, this is "the famous crossing curve of species immigration and extinction." One curve (species coming in) descends, upper left to lower right, and flattens as empty islands fill up with plants and animals. Gradually there's less room for new arrivals from elsewhere. The other curve (species disappearing) rises from lower left to upper right and gets steeper as existing populations die out. The curves, immigration and extinction, meet in the middle.

This intersection, an equilibrium point, is the carrying capacity of an island, determined by its size and isolation (distance from the mainland and from nearby islands). MacArthur and Wilson call it a dynamic equilibrium because species turnover continues; what's constant throughout the flux is the number of species: more on bigger islands. This was just as Darlington predicted— and in the same "strikingly orderly" ratios, noted MacArthur and Wilson. Only now this seemed like a rule to count on, not simply a rule of thumb. Wilson wrote, "The number of species (birds, reptiles, grasses) approximately doubles with every tenfold increase in area."

Forty years after publication, Wilson explained that one reason he and MacArthur wrote their book was to disturb the equilibrium in their own field of biology, where MacArthur's interest was birds and Wilson's was ants. Both sensed that the work of naturalists, such as themselves, was getting shouldered aside, with university positions and grants going, he wrote, "to the newly triumphant emergence of molecular biology" following the headline-grabbing reveal of the double-helix structure of DNA. So, Wilson went on, "we soon narrowed our conversations down to the following question: How could our seemingly old-fashioned subjects achieve new intellectual rigor and originality compared to molecular biology?"

The Theory of Island Biogeography was a spectacularly successful answer. It's led to innumerable research projects and, biologist Theodore H. Fleming noted, quickly made "island thinking" part of the "conceptual toolkits" of generations of ecologists, evolutionary biologists, and conservation biologists. It's been compared to Darwin's *On the Origin of Species* in terms of how often it's cited. Along the way there have been challenges to the theory—it did not take into account the ways extinction rates accelerate when people arrive and settle on an island, for example.

But what gives it lasting appeal is that by investigating islands, which make up a little more than 3 percent of all the land on earth, and using what MacArthur and Wilson called "simplifying models of complex phenomena," you can get closer to answering even older questions: Why are plants and animals found where they are, and where did they come from? Only two hundred years earlier, Carl Linnaeus, the eighteenth-century Swedish botanist and zoologist who created the orderly process for naming and counting species, thought it logical and rational that all living things came from a "Paradisical Mountain" on the equator, and that after the biblical flood, plants and animals migrated down the mountain where Noah's Ark landed.

The "island thinking" of MacArthur and Wilson—there's a stir of poetry about it. Not "no man is an island," but the idea that almost anything can be seen as an island. In a paper called "Flowers as Islands," three Stanford biologists looked at the microfungi in nectar in sticky-monkey flowers favored by hummingbirds, with the premise that "flowers are discrete, island-like habitats," and they were pleased to find an orderly, nonrandom pattern to the dispersal of the microbes.

Islands in the landscape

In *The Theory of Island Biogeography,* island thinking, as it might apply to landscapes, is right there in chapter 1, even before the graph of the crossing curves. There's a brief discussion of the "insular nature" of streams, caves, tide pools, and tundra, and a warning that when a species becomes islanded, so to speak, it has a harder time recovering from floods or disease. The same principles apply, write MacArthur and Wilson, and "will apply to an accelerating extent in the future, to formerly continuous natural habitats now being broken up by the encroachment of civilization." To illustrate how a forest in Wisconsin got islanded, a map shows the woods gradually disappearing until only scattered remnants remain. Then the book moves on and falls silent on this subject. Even so, it inspired others to pick up on the idea.

People who wanted to preserve isolated mountaintops in the Southwest started calling them Sky Islands—a rising tide of urban sprawl was making it difficult for wide-ranging animals like black bears to reach the high peaks. Wilson told science historian Fred Powledge that, in the 1960s, neither he nor MacArthur thought much about how their theory applied to conservation. It was only at a 1972 memorial symposium for MacArthur that Wilson first put forward the notion that any nature reserve was a marooned landscape "destined to become an island in a sea of habitats modified by man."

The species-area math cut two ways: more land set aside meant protection for more species, but take away the vast expanse of a landscape and only a fraction of its plants and animals would survive.

For Wilson this was a first step toward Half Earth thinking—when animals can leave a place but can't get back in, constriction itself becomes a prediction, undermining equilibrium so effectively that that place automatically starts losing species. "It's common sense, when you think about it," he told Powledge. This was dramatically confirmed in 1987 by an article about parks in the Science Times section of *The New York Times* (which shared the page with a piece on a new atom smasher said to create "explosions so violent that they mimic the state of the universe when it was only a fraction of a second old"). The parks article lede was just as alarming: "Many species of mammals are disappearing from North America's national parks solely because the parks—even those covering hundreds of thousands of acres—are too small to support them."

The data came from William D. Newmark, an ecologist who surveyed fourteen national parks in the United States and Canada (parks keep good records) only to find what the *Times* called "a striking loss of species in the great Western parks, from grizzly bear to red fox to white-tailed jackrabbit, on a scale unrecognized by wildlife officials. Parks as vast as Yosemite and Mount Rainier have lost more than one-fourth of the species originally

Species Vanishing From Many Parks

By JAMES GLEICK

MANY species of mammals are disappearing from North America's national parks solely because the parks — even those covering hundreds of thousands of acres — are too small to support them.

New research has found a striking loss of species in the great Western parks, from grizzly bear to red fox to white-tailed jackrabbit, on a scale unrecognized by wildlife officials. Parks as vast as Yosemite and Mount Rainier have lost more than one-fourth of the species originally found there, and smaller parks have lost as many as 35 to 40 percent.

As roads, housing development and deforestation take hold around park boundaries, they isolate animal populations in regions that seemed like spacious havens when the parks were established 70 to 90 years ago. By then some species had already been killed off or driven out of the park areas.

But the first complete survey of major mammal populations throughout the Western parks, published in the current issue of Nature, shows that the loss has not abated.

"It's extraordinarily important information," said Michael Soulé, a research fellow at the National Zoological Park in Washington and the president of the Society for Conservation Biology. "The scale is beyond anything that people had appreciated or feared, and it's bound to have an important impact on the management of the national park system and of wildlands in general in the United States."

Although individual park managers and scientists have noted some of the local extinctions, few appreciated a trend that took shape over the better part of a century.

A Gloomy Signal on Survival

Survey of mammals in Western parks, including Mt. Rainier, pictured, describes loss 'beyond anything that people had appreciated.'

Name of park (Area in square kilometers; age of park)	Mammal species lost since establishment of park	Proportion lost of original resident species*
Bryce Canyon (144 sq. km; 61 years)	5	36%
Mount Rainier (976 sq. km; 85 years)	7	32%
Rocky Mountain (1,049 sq. km; 69 years)	2	31%
Yosemite (2,083 sq. km; 94 years)	4	25%
Grand Teton-Yellowstone (10,328 sq. km; 83.5 years)	1	4%
Lassen Volcanic (426 sq. km; 77 years)	6	43%
Kootenay-Banff-Jasper-Yoho (20,736 sq. km; 84.5 years)	0	0%

*Includes losses before and after park establishment.

Photo by Rapho/Esther Henderson; Source for chart: Nature; The New York Times/Feb. 3, 1986

From The New York Times, *1987*

found there, and smaller parks have lost as many as 35 to 40 percent." All a result of constriction, though Newmark used the word "insularized." From the *Times:* "As roads, housing development and deforestation take hold around park boundaries, they isolate animal populations in regions that seemed like spacious havens when the parks were established 70 to 90 years ago."

This sudden, stark realization about the inadequacy of national parks got people to think bigger, at the scale of the continent. But could this understanding have come any sooner? In 1952, fifteen years before *The Theory of Island Biogeography,* Eugene G. Munroe, a Canadian entomologist who was considered *"the* lepidopterist" (butterfly and moth specialist) as well as the world authority on a superfamily of small moths with at least sixteen thousand species—he's reported to have collected his first moths at age four—independently discovered the equilibrium theory of island biogeography. His 555-page Cornell PhD dissertation on the subject was never published, but decades later lepidopterist Paul Feeny happened across it in the university library.

Biologists found Munroe's insights remarkable, and in nearly all aspects his equilibrium theory was equivalent to MacArthur's and Wilson's. When people asked Munroe why he never pushed to get his work recognized, he wrote to biologist James H. Brown that he had "competing interests and pressures." Munroe seems to have been a lot like Philip Darlington, whom he turned to for advice when writing his dissertation. In Munroe's obituary, a colleague described collaborating with him as a great joy, and mentioned his "modesty and his preference to praise the work of others and/or talk about moths and butterflies."

In 2018, two Norwegian and two Czech scientists published a paper in the *Journal of Biogeography* intended to restore the legacy of the "Stockholm group": five Nordic botanists—a Finn and four Swedes, one of whom later received a Nobel Prize in chemistry. In the 1920s the Stockholm group also wrote about the species-area relationship, came up with their own math, and then "faded into oblivion," noted the writers of the paper. The Stockholm group published, but mostly in Swedish and Danish, "buried in the wrong journals in inaccessible languages" at a time when "the world was much less interconnected." The paper concluded that the scientific community just wasn't ready for the Stockholm group, and only now could they get the credit they deserve.

A community to which we belong

"An Armageddon is approaching," E. O. Wilson wrote nearly two decades ago. "Not the cosmic war and fiery collapse of mankind foretold in scripture. It is the wreckage of the planet by an exuberantly plentiful and ingenious humanity." Maybe the shadow of a sixth biological annihilation—now often simply called "6"—fully awakened the science of 50 percent, making it a kind of night-blooming plant.

Long before we could count to "6," the British naturalist Alfred Russel Wallace, a creator of the theory of evolution (independently of Darwin), warned against ignoring extinction in an 1863 paper, "On the Physical Geography of the Malay Archipelago":

If this is not done, future ages will certainly look back upon us as a people so immersed in the pursuit of wealth as to be blind to higher considerations. They will charge us with having culpably allowed the destruction of some of those records of Creation which we had it in our power to preserve; and while professing to regard every living thing as the direct handiwork and best evidence of a Creator, yet, with a strange inconsistency, seeing many of them perish irrevocably from the face of the earth, uncared for and unknown.

In 1985, Michael E. Soulé and several other American biologists founded the Society for Conservation Biology. At first dismissed as "an odd assortment of academics, zookeepers and wildlife conservationists," it now has four thousand members around the world. Conservation biology is called a "discipline with a deadline"; Soulé's term is "crisis discipline." Conservation biology, he wrote in *BioScience* in 1985, was as much a rallying point as it was a new field, studying life not just to study it, but to save it: "Its relation to biology, particularly ecology, is analogous to that of surgery to physiology and war to political science."

Another founder, David Ehrenfeld, author of *Becoming Good*

Ancestors, wrote in a 1992 letter to *Science* magazine, "Conservation biology is not defined by a discipline but by its goal—to halt or repair the undeniable, massive damage that is being done to ecosystems, species, and the relationships of humans to the environment." Urgency was the only possible response to an ongoing emergency. As Ehrenfeld put it, "Many specialists in a host of fields find it difficult, even hypocritical, to continue business as usual, blinders firmly in place, in a world that is falling apart." His words echoed those of Aldo Leopold, the conservationist and philosopher whose *A Sand County Almanac* has sold 2 million copies in fourteen languages. A generation before Ehrenfeld, Leopold called the planet "a world of wounds" that "believes itself well and does not want to be told otherwise."

"To safeguard biological diversity, larger-scale and longer-term thinking and planning had to take hold," Michael Soulé, Reed Noss, and a third colleague, Curt Meine, wrote on the twentieth anniversary of the founding of the Society for Conservation Biology. Meine is Leopold's biographer, and Leopold in the 1940s predicted that a "land ethic" would emerge one day, writing that "when we see land as a community to which we belong, we may begin to use it with love and respect." The emergency could accelerate a different kind of evolution, even over the course of a single lifetime.

Indeed. "Over the years I have evolved from biologist to conservation biologist," George B. Schaller has noted. He's the National Book Award–winning author who helped give Dian Fossey her start, contributed to the survey of the 19-million-acre Arctic National Wildlife Refuge in Alaska, stopped the slaughter of giant pandas, and has worked to set up more than twenty protected areas and parks around the planet. This includes a proposal for an International Peace Park in the remote, rugged sliver of mountains where Afghanistan, Pakistan, Tajikistan, and

China converge, and the Chang Tang Nature Reserve in Tibet, the centerpiece of more than 122 million acres of protected high grasslands that are home to snow leopards and some of the last herds of wild yaks and Tibetan chiru, an antelope thought to have the finest wool in the world. Schaller is sometimes described as a natural wonder of the world himself—photos of his fieldwork show him canoeing in Alaska in 1952 with a pet raven, and on a boat in Brazil in the 1970s cradling his pet white-lipped peccary, a South American wild pig.

"Research is easy; conservation most definitely is not," Schaller writes in *A Naturalist and Other Beasts,* one of his nearly twenty books. Caution has become his trademark. "Remember, there are no victories in conservation," he says. "You may have a temporary 'looks good.' Suddenly things change and the fight begins all over. So, we've got to continue fighting." In 2017, sixty-one years after Schaller participated in the survey that led to its protection, Congress authorized drilling for oil and gas in the Arctic National Wildlife Refuge.

Corridors

Reed Noss told me he vividly remembers getting laughed out of Ohio. That was in 1983, before his plan for Florida wildlife—which never would've existed if he hadn't left Ohio. The Ohio plan, a network of reserves, would have protected 3.6 million acres, much less than the 25-million-acre Florida plan. And it's the Florida plan that's considered the foundation for thinking of grander geographies. It's also a template for what's called "vision mapping" by the Wildlands Network, which Soulé and Noss cofounded in 1991, with Dave Foreman, cofounder of Earth First!, and Douglas Tompkins, cofounder of two clothing companies, The North Face and Esprit. This led to Soulé's Spine of the Continent (also called the Western Wildway), an effort to protect the Rocky Mountains from Mexico to the Yukon.

For Noss, thinking big is a matter of space, time, and what he calls ambition. "Space and time have to do with looking farther afield and also further back than you might've expected to," he said. "Ambition is about what's possible and making sure it happens. We never get everything we ask for, but it's logical that if you ask for something really big, you get closer to it after a compromise than if you don't ask in the first place."

He told me that thinking big came to him naturally, growing up in Ohio. It's a state rich with fossils from the Ordovician-Silurian extinction 440 million years ago, the first mass extinction. In childhood, Noss got a sense of hundreds of millions of years—and of the shadow. Things like trilobites could disappear, never to return. More recently, maybe 30 million years ago, southern Ohio became part of the mixed mesophytic forest. It is one of the world's oldest and most biologically rich temperate-zone hardwood forests, and originally spread over three continents. Now virtually gone in Europe, though still part of central China and the eastern United States, this forest is said to offer "a rare glimpse of what life was like in the ancient forests."

François André Michaux, a French botanist, saw these American woods in their full glory when he canoed down the Ohio River in 1802, through the tumbled ridges, coves, and hollows at the edge of Appalachia: "In more than a thousand leagues of the country, over which I have traveled at different epochs, in North America, I do not remember having seen one to compare with the [Ohio Valley] for the vegetative strength of the forests." Michaux, then in his thirties, came to America at fifteen with his father, a botanist sent by the French king on a mission to regrow forests in France that had been stripped to construct warships. His father sent home sixty thousand living plants and ninety boxes of seeds.

In Michaux's day, the Ohio hardwood forest had white oak trees four feet across that, Michaux said, "had a straight trunk *without a single branch* for seventy feet." This forest started slipping away, stripped not by farmers—the soil was too poor—but for the huge charcoal furnaces of the iron industry. By the 1900s, after the ironworkers left, the recovering forest found its champions in "two lady doctors," as people in the area called them.

Ohioans, as it happened: E. Lucy Braun, a plant ecologist, and her older sister and assistant, Annette, an entomologist. Both earned PhDs, lived together in Cincinnati throughout their lives, and did research by driving 65,000 miles in a car that Lucy Braun bought in 1930. They could, of course, finish each other's sentences. A young Noss heard much about Lucy Braun early in his career and met some of her students at the University of Cincinnati. Her masterpiece, *Deciduous Forests of Eastern North America* (1950), has been called a textbook that reads like a novel. While most temperate forests are dominated by two or three species of trees, Lucy Braun found eighty in southern Ohio's mixed mesophytic forest (she coined this term). Sheltered from the ice ages, its hollows are, she wrote, "the likely ancestral source of most temperate-zone forest species in the eastern United States."

The Ohio Valley forest was coming back, but it had lost its wide-ranging species—black bear, bobcat, river otter. They weren't gone for good, merely "extirpated," no longer present in the area. Some species, though, were gone for good. Noss remembers, when he was twenty, volunteering on an archeological dig and excavating the skull of a passenger pigeon, a bird that only a century earlier had numbered in the billions. At that moment, "extinction became very personal," he told me, and since then

E. Lucy Braun

it's been his life's work to shine a light into the shadow to "get every species back in its rightful place." It's a process now called "rewilding."

"It was my dream," Noss says, "to restore this whole deciduous forest ecosystem and re-create wilderness in the Ohio Valley." He thought he'd found his chance working for the Ohio Department of Natural Resources when his boss asked him to come up with a conservation project. The state had already done right by one of the astonishing features of the woods—the "prairie peninsula," pockets of Ohio grasslands three hundred miles east of Illinois (the Prairie State). What was missing was a strategy for the far larger and equally extraordinary "mother forest."

Thinking big meant starting small. In the 3.6 million acres of the mother forest, there were a million acres already protected to some degree—a dozen state and national forests, and Nature Conservancy preserves—and in between, most of the privately owned land was still only lightly settled. Add on just a fraction of land—the right fraction, land that connects—and the area could again become the kind of deep wilderness François André Michaux would recognize.

The way I think of it, it's like a bunch of isolated cabins that might not be big enough for a whole family but, if joined up by passageways, these cabins could become a single, many-roomed house, a house of the right size and shape. It would have rooms large enough to assure an ecosystem's stability and long passage-ways wide enough for animals to roam.

Cores and corridors, Noss called this. "Corridors"—not the world's most inviting word. It sounds restrictive and unwelcom-ing, institutional, too much like hospital or school hallways, places for hurrying along or skulking through on the way to where you're really going. But "corridors" was a word biologists were already comfortable with, so it stuck. To change images again, jewels could be made more valuable by creating a bracelet or necklace out of separate gemstones. Even in a crowded state like Ohio, all this was still possible. The links were clear. The gems were there.

If and when, however. *If* people throughout the area welcomed the idea of living in the mother forest; *when* they sensed that if lost again it would be irreplaceable this time. Noss thought the pulling together might need a generation to take hold, though there was a way to get it started, a first link. You could take The Edge, more formally the Edge of Appalachia, a 20,000-acre Nature Conservancy forest-and-prairie preserve; at the heart of it was the 42-acre Lynx Prairie Lucy Braun found on her trav-els. Then connect The Edge to a nearly 64,000-acre state forest, Ohio's largest, a few miles away, nicknamed the Little Smokies of

Ohio because of the dense bluish fog that clings to its peaks. "An unparalleled opportunity," Noss called it.

Then came the laughter. "Outrageous," "audacious," "way too big," "politically disastrous if it comes from us," said the higher-ups in his agency, though his own boss actually liked the idea. "Way too ambitious," said state chapters of several environmental groups, though not The Nature Conservancy, which since the 1980s has been, as they say, "patchworking together" The Edge and the state forest by protecting a 6,000-acre Sunshine Corridor along a ridge between the two. But Noss left the state, taking with him an understanding of how to assemble and unite the natural workings of a place by thinking big about the ecosystems that can't move around and the animals that have to.

Everything Noss saw as a possibility in the Ohio Valley remains possible; the area's still forested and a bit remote. "It's amazing," Noss says now, "in the last few years, fifty percent—more than I was reaching for back then—has been catching on as reasonable." In 1992, Noss was the first to go public with this percentage, writing in *Wild Earth* magazine that, to protect ecological integrity, "at least half of the land area of the 48 conterminous states" should "within the next few decades" become core reserves and corridors. In effect this was a "50 by 2020" declaration.

Reception to the idea, on a national level, was very much what Ohio's had been—despite Noss introducing 50 percent carefully, explaining that wilderness recovery was a "thorny" issue and that each region had to be treated individually. "People," he says, "laughed or got angry, saying this was ridiculous and too radical and would discredit the whole conservation movement." It probably didn't help that in the same piece Noss said that if the U.S. population ever declined, a goal could be to keep at least 95 percent of a region managed as wilderness, surrounded by lightly inhabited wildlands.

Rethinking consciousness

In 1988, the same year the word "biodiversity" (without the capital B and D) appeared in print for the first time, Norman Myers, a British ecologist, introduced new terms—"hotspot areas" or "hotspots." Applying species-area math to twenty years of tropical forest destruction on three continents, areas assumed to contain at least half the earth's species, Myers saw a "major extinction spasm impending" and an urgent "super-imperative of stemming" it.

Myers had an unusual insight into a solution; in fact, he took pride in "thinking sideways," as he said. He grew up on a sheep farm in the north of England, no electricity or gas, became a colonial administrator in Kenya in the last days of the British Empire, and stayed on as a high-school teacher and then a wildlife photographer, only later going to grad school. His approach to biocide: start little to go big. Within the species-rich forests are smaller areas that are even species richer, full of endemics—plants and animals found nowhere else. Biodiversity hotspots. They cover 2.3 percent of the land but support more than half the world's plants and shelter 43 percent of birds, mammals, reptiles, and amphibians. Save them first, Myers said.

"The hotspots thesis," he said, "has the potential to reduce the mass extinction underway by a whopping one-third." He called this the "clincher factor," and was particularly pleased that E. O. Wilson called the idea "the most important contribution to conservation biology of the last century." Originally Myers came up with ten hotspots; now there are thirty-six—no longer "spots" or only in the tropics. Number 36, the North American Coastal Plain, or NACP (Noss led the team that got it designated in 2016), covers 280 million acres up the East Coast from Mexico through twenty-one states. The Delmarva Peninsula's in it; so are Long

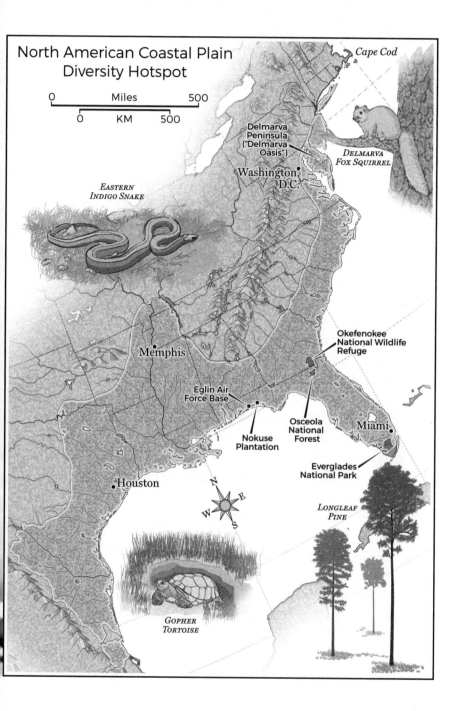

North American Coastal Plain Diversity Hotspot

0 — Miles — 500

0 — KM — 500

Cape Cod

Delmarva
Peninsula
("Delmarva
Oasis")

*DELMARVA
FOX SQUIRREL*

Washington,
D.C.

*EASTERN
INDIGO SNAKE*

Okefenokee
National Wildlife
Refuge

Memphis

Eglin Air
Force Base

Osceola
National
Forest

Miami

Nokuse
Plantation

Everglades
National Park

Houston

N
W · E
S

*LONGLEAF
PINE*

*GOPHER
TORTOISE*

Island, Cape Cod, Martha's Vineyard, and Nantucket. So for that matter are cities like Houston, New Orleans, Miami, Savannah, Charleston, Philadelphia, and, because Long Island is included, Brooklyn and Queens.

Like the first thirty-five hotspots, the coastal plain is a nick-of-time landscape, a place of imperiled profusion, home to a wealth of species found nowhere else that could be swept aside within a generation. According to strict criteria set up by Myers and several colleagues, each hotspot has to have more than 1,500 native plants and must already have lost at least 70 percent of its original habitat—and in the coastal plain, 86 percent is gone.

"Sometimes," Noss told me, "something precious is right under our noses but goes unnoticed. The North American Coastal Plain was a hotspot hidden in plain sight. Lots of people find it ordinary, monotonous low-lying countryside. Most hotspots are in more mountainous terrain—the assumption was you needed all kinds of variation in climate and topography for large numbers of native species to coexist. Some botanists and zoologists knew about the proliferation of life in the coastal plain, but nobody had gotten around to tabulating it. So I made a deal with them—if they compile the lists, I'll take the lead on writing the paper. The totals surprised all of us: 1,816 species of plants found only in the NACP, along with 51 species and subspecies of birds and 114 species and subspecies of mammals. Other lists show it's also a hotspot for turtles, ants, grasshoppers, and lichen. Not to mention a hotspot for people too, with more and more moving in even while sea levels are rising."

The coastal plain can also help with global warming. Here's Noss again: "Many NACP species are fire dependent, having adapted to millions of years of lightning strikes. In Florida, there are ancient grasslands, themselves millions of years old, that have become underground forests in effect, sending up only a few shoots that fires can burn off, hiding their massive trunks and branches belowground, branches that if aboveground would burn and add to the carbon dioxide warming the planet." Underground forests, like overhead rivers—topsy-turvy aspects of the biosphere just beyond the edge of sight. "Hey," Noss went on. "Just another reason to think big."

These forests are part of the "new botany," a turning away from Aristotle's idea that plants don't amount to much, still too common an assumption. Aristotle's Ladder of Life, 2,300 years old, put humans at the top and plants near the very bottom— below every animal, even stay-put "plant animals" like corals and sponges. Plants were only one rung above lifeless minerals, whose only purpose, Aristotle thought, was to come to rest. In *The Cabaret of Plants*, British nature writer Richard Mabey says that plants "have come to be seen as the furniture of the planet, necessary, useful, attractive, but 'just there,' passively vegetating. They are certainly not regarded as 'beings' in the sense that animals are." Adrian Barnett, an ecologist, writes in *New Scientist*, "They seem too slow, and too different, to register as intelligent . . . a form of mechanistic half-life of simple growth and response— a flatlined existence devoid of subtlety, strategy and learning."

But among the new botanists there's the understanding that, for hundreds of millions of years, the trees in forests have been acting like a community, or a superorganism, aware of one another, warning of insect attacks, exchanging nutrients, and sending supplies to seedlings, connected at their tangled roots by miles of fungal fibers nicknamed the "Wood Wide Web."

"Intelligence is a term fraught with difficulties in definitions," cautions Anthony Trewavas, a plant physiologist. In *New Scientist*, Stefano Mancuso, a plant neurobiologist, says that because of "a kind of brain chauvinism," people "think that a brain is something that is absolutely needed to have intelligence." In the same

The Wood Wide Web

article, Michael Marder, a Canadian who teaches in Spain, is called "the lone plant philosopher for now." Marder says it's time for us to think about attention and consciousness as not uniquely human: "I want to rethink the concept of intelligence in such a way that human intelligence, plant intelligence, and animal intelligence are different sub-species of that broader concept."

Common cause

Taking a phrase coined in the business world, 50 by '50 is a BHAG—a big hairy audacious goal, a commitment to a galvanizing project that has a 50 to 70 percent chance of success and may take up to thirty years to achieve. "You are no longer managing for the quarter but for the quarter century," says Jim Collins, co-inventor with Jerry Porras of the term. Why 50 to 70 percent? There's no urgency when there's a 100 percent chance of crossing the finish line, and too little hope with a 10 percent chance. In an *Inc.* magazine interview, Collins explains that BHAGs (pronounced *bee-hags*) can change the world, even if the effort falls short: "And in the end, what might happen is it's an inspiration to the next generation who will pick it up from there." Examples he cites include President Kennedy's promise to put a man on the moon within a decade and Bill Gates's pledge to put a computer on every desk and in every home.

Saving 50 percent of the land by 2050 to protect up to 90 percent of the species may not be the ultimate BHAG. Some biologists look beyond to an even more audacious one. Noss, for instance, told me it's "quite possible to maintain *all* the original species in, say, twenty-five to seventy-five percent of an area—if the reserves are carefully selected and connected."

But after you work out the numbers, there's something not quite calculable, something beyond the spreadsheets and off the charts. Such as, say, the coordinated thinking of the largest organism on the planet—47,000 genetically identical quaking aspen trees, all stemming from a single root system and ranging across 107 acres in Utah's Fishlake National Forest.

There are also deep-seated tendencies of the human mind, given names by biologists only recently. In 1984 E. O. Wilson said people have an innate urge to explore and affiliate with the rest of life: "From infancy we concentrate happily on ourselves and other organisms. We learn to distinguish life from the inanimate and move toward it like moths to a porch light. Novelty and diversity are particularly esteemed." He called this idea the "biophilia hypothesis" and wrote about it in the book *Biophilia*. "To an extent still undervalued in philosophy and religion," he said, "our existence depends on this propensity, our spirit is woven from it, hope rises on its currents."

A near-universal trait was uncovered in 1971 by an English epidemiologist, J. Ralph Audy. He had a PhD and an MD and spent much of his life in the field in Africa, India, and Southeast Asia, where, according to a memorial tribute in the *Journal of Medical Entomology*, he was prized for his "inexhaustible supply of jokes, stories, limericks, songs and games" while discovering "a plethora of new viruses." He was just old enough to have started out as a cavalry officer in what in 1940 was still the British Army's Somaliland Camel Corps, and in his final years was a foundation director and professor in San Francisco. In several essays, Audy wrote that almost every species approaches the world as an imperfect, only roughly suitable place, a fixer-upper, and sets about modifying it to make life safer and more comfortable for itself.

With lesser or greater effects. In some cases it literally means feathering your own nest; many bird nests are made out of a few twigs, some dried grasses and feathers, and perhaps spiderwebs. Other alterations can change the lives and circumstances of animals and plants for miles around. Beaver dams, built from trees up to three feet thick, can create vast wetlands. The largest beaver dam in the world, in Wood Buffalo National Park, in Alberta, Canada, is half a mile long.

Audy called all these constructed habitats—nests, dams, cocoons, burrows, cars, skyscrapers, hospitals—"ipsefacts," meaning "things they make themselves." People are the most ingenious ipsefactors: creating pleasant temperatures year-round; light at night; warm and cool springs that at the touch of a faucet flow through houses. But our "niche constructions" (another term for it) have gone so far toward rebuilding the planet in our own image, we've made life more difficult for other creatures and their ipsefacts, and along the way have walled ourselves off from all but a faint awareness of the living world beyond.

50 by '50 thinking is about amplifying that awareness, linking the selflessness of biophilia with our self-centric nature as ipsefactors. We can, for instance, try to listen to what other species are telling us.

The Animal as Storyteller

On June 6, 1991, a mile up in the Canadian Rockies near Banff National Park, a drenching storm turned the fur of a gray wolf black. A poetic zoologist, who'd just captured this five-year-old female wolf to fit a radio transmitter around her neck, named her Pluie ("rain" in French) and then let her go. It seemed like routine research but was anything but. It was, as they'd discover, a profound glimpse into a larger geography where animals are the guides, scouts, surveyors, and mapmakers.

Pluie wore something unprecedented that now, in a world flooded with 3.5 billion smartphones capable of pinpointing people's locations moment by moment, seems almost quaint. She was one of the first wolves, and actually one of the first mammals, to wear a collar whose pinging signal could be picked up by satellites flying 530 miles above the earth and then sent back to a receiver on the ground.

This was shortly before the Defense Department's Global Positioning System (GPS) was made available for civilian use, as part of a peace dividend that later made smartphones possible. GPS—

twenty-four satellites 12,600 miles overhead, launched during the Cold War—kept track of American submarines and hostile missiles. More expensive but more accurate than earlier satellites, GPS made possible animal tracking based on a vastly tall, invisible isosceles triangle of signals and relays.

It was known that wolves wander. Packs roam, and individuals peel off, looking for food, a mate, or another pack. In a couple of years Pluie, considered a loner, might log sixty or seventy miles in back-and-forthing. For six months Pluie told her researchers exactly that: she moved a bit north and briefly joined a Banff wolf pack. Then came two surprises, as wildlife biologist Karsten Heuer, one of the trackers, recounts in his book, *Walking the Big Wild*.

The first came in late November, when transmissions stopped. Researchers checked all nearby roads and hiked ridge after ridge. No Pluie. Zoologists closed the books on her, figuring either the wolf or the battery must have died. A month later came the second and much bigger surprise, a real jolt. A civilian branch of NASA got in touch to say a satellite had picked up Pluie's signal hundreds of miles to the southeast, down below the U.S. border in Montana. Pluie hadn't stopped broadcasting for a moment. The researchers hadn't thought beyond their expectations and so weren't looking nearly far enough away.

Over the next eighteen months, before transmissions ceased once more, zoologists followed Pluie's every move and with growing amazement. At one point during her Montana trek, she was walking through open prairie, east of Glacier National Park. She then turned west for hundreds of miles, went up and over the Rockies, skirted Glacier's southern edge, crossed planted fields and busy highways, traversed rivers and lakeshores in the Idaho Panhandle, passed by Coeur d'Alene, a resort town Barbara Walters once called a little slice of heaven, and got all the way to

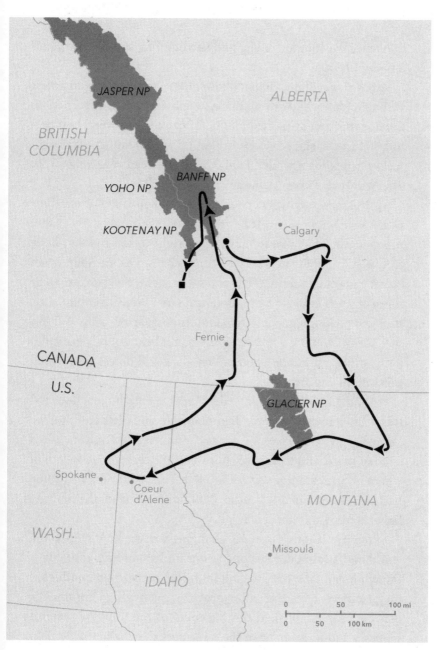

Pluie's wanderings

Spokane, Washington, a city and surrounding suburbs with half a million people.

After that, Pluie changed direction again, doubling back through Idaho and Montana, and moving north returned to Canada through British Columbia. After another several hundred miles of travel, she was back in Banff. "We thought she was on a pickup truck for a while," Paul Paquet, the lead researcher, told *The New York Times*, "she was moving so fast."

Wolves are somewhat nearsighted, but in other ways their world is far more vivid than ours, as Wolf Country, a website, makes clear. They like to move at an eight-to-ten-mile-an-hour trot, a speed that increases to thirty-five miles an hour when chasing another animal. They can see at night, and because their sense of smell is about a hundred times more sensitive than ours, they can pick up a scent a mile and three-quarters away. Within the Rocky Mountain forests, Pluie could hear sounds six miles away, while out on the open prairie she could detect sounds ten miles off in any direction.

In late 1993 Pluie went silent again. Her bullet-damaged transmitter turned up on a mountainside outside Banff, but we know she kept on roaming for two more years. Hunters killed her, her mate, and their three pups in 1995, eighty miles north of where her transmitter was found. But for a year and a half Pluie had been a kind of 24/7 radio station, an all-news-all-the-time bulletin from another species.

Some inventions resonate, changing forever how we see the world and adding dimensions to our understanding of reality—though it can take time to sink in. In the early seventeenth century, Galileo's telescope, after provoking convulsive arguments, convincingly established the hugeness of the universe. Microscopes, an innovation from the same period, opened the domain of the minuscule and two centuries later tightly linked human health to the existence of tiny living things with the rise of the germ theory of disease.

It took quite a mash-up of recent inventions—radios and satellites and software—to give Pluie a voice. More than a quarter century after she went off the air, conservationists and ecologists, landowners and developers, and vacationers and government officials are still coming to terms with her messages. Far from being static, opaque entities lurking in our peripheral vision, the landscapes around us are alive with purpose, drama, movement, a constant parallel presence that directly intersects with our lives. Landscapes are verbs, not nouns. The living world operates at an epic scale that dwarfs the traditional conservation thinking of the last 150 years.

To keep 90 percent of species alive, the 50 by '50 goal, we have to know where and how these species are actually living their lives. Pluie's irregular path, something between a squashed circle and a lumpy triangle, encompassed a forty-thousand-square-mile territory not indicated on any map because it ignored political and geographic boundaries. She was a transnational wolf, crossing parts of three U.S. states and two Canadian provinces as she made her way around an area ten times that of Yellowstone. Because places stay put but animals don't.

The land revealed the answer

In the 1830s, an artist, George Catlin, visited forty-eight Great Plains tribes and painted stark and striking portraits of their great chiefs. Catlin warned that the Great Plains would be swept away unless a "magnificent park" was created "by some great protecting policy of government"—"a nation's Park, containing man and beast, in all the wild and freshness of their nature's beauty!"

It took forty years and needed an act of Congress before Yellowstone, the world's first national park, came into being in 1872.

In the 1990s, Harvey Locke, a young Canadian lawyer and conservationist, came up with his own ideas about protecting land—and didn't want to wait decades. His family has lived in or near Banff National Park for seven generations; his great-great-grandparents arrived there in the 1870s after a steamboat they were on was forced to stop for eight hours to allow a buffalo herd to pass. In the years Pluie was still wandering, Locke got swept up in what he called his "first deep brush with biology." At a six-week-long environmental hearing in 1992, witness after witness—people Locke had lined up, experts on wolves, grizzly bears, and bighorn sheep—demonstrated how a golf course and resort village planned for a seemingly empty stretch of forest south of Banff would in fact bring an irreparable halt to the movement of large animals that came surging through the place, an unseen but still a permanent presence (unlike buffalo, which were long gone). A permit for the golf course and resort village was denied.

The next summer Locke rode on horseback into the heart of the M-K, the Muskwa-Kechika, sometimes referred to as North America's Serengeti and the continent's biggest well-kept secret. It's a British Columbia wilderness seven times the size of Yellowstone. After a day seeing elk, eagles, caribou, and moose, and a grizzly bear track and a wolf track side by side, Locke sat around a campfire, trying to think about what spaces animals might need.

Already it wasn't just Pluie—one collared wolf traveled from Glacier National Park in Montana down to Yellowstone, and another went from Glacier almost up to Alaska. In yet another unexpected find, Peter Sherrington, a Canadian oil-patch geologist, was out bird-watching in early spring 1992, trying to catch sight of a pine grosbeak, a plump pinkish-red finch. Moving his binoculars around, Sherrington noticed a small dot high in the sky, which turned out to be a golden eagle, a bird with a seven-foot wingspan and considered far more of a rarity. Then he saw two more eagles, and after that another, and another. By the end of the day he and a friend spotted 103—and discovered the golden eagle highway, an eleven-thousand-year-old migration route along the Rockies that thousands of golden eagles use to travel between Mexico and Alaska, soaring and gliding at ten thousand feet and higher.

At the M-K campfire that night, as Locke later said, "the land revealed the answer." A week before, he'd hiked in another wilderness, miles to the south. And yet now, as the stars glittered above, he felt he had stayed in one place, still deep within, as he put it, "one gigantic linear ecosystem." A place without a name, a Pluie-sized unit of territory—bigger than a U.S. state or a Canadian province, smaller than the continent, a concept people could grasp. "The scale that really matters is what's going on in people's heads every day," he said. This was, for Locke, the way to match the land to the way people could understand it.

Y2Y

In a 2005 speech to a National Park Service conference, Locke talked about *American Progress,* an iconic, often reproduced image of the settling of the American West. The painting, by John Gast, dates from 1872—the same year Yellowstone became a park—and filled American imaginations for decades. It shows Native Americans, a buffalo herd, an elk, and a bear retreating westward under darkened skies while miners and farmers (all white men) advance from the east, where a brilliant sun is rising. Hovering over the scene is a golden-haired young woman—"Progress"—described by the travel-guide publisher who commissioned the painting as "beautiful" and "charming."

Scantily clad in a diaphanous gown, "Progress" has a Star of Empire on her forehead and a schoolbook in one hand, while her other hand trails telegraph wire that stretches back east on a long line of poles. Stagecoaches and railroads follow her. It's technology in service of mass eviction, of people and animals alike. But newer technology, like Pluie's collar, which amplifies the signals from an animal's life, could serve the opposite purpose and be "very, very powerful," Locke said, "because it tells us what to do."

At the campfire, Locke pulled a topo map from his backpack and began to write in the margins. What came into focus for him was something he'd been rolling around in his mind—

American Progress, *1872*

"Yellowstone to Yukon." A name that lent itself to a catchy acronym, Y2Y, and had policy and poetry at either end: Yellowstone as one of the world's first experiments with protecting nature, and Yukon because it evokes wilderness—"the freshness, the freedom, the farness" of a wild frontier. Which is a line from the jingly, haunting poem "The Spell of the Yukon," by Robert W. Service, a cowboy in the Yukon Territory a century back. The poem ends: *"It's the great, big, broad land 'way up yonder / It's the forests where silence has lease; / It's the beauty that thrills me with wonder, / It's the stillness that fills me with peace."*

Locke's scribbled-down notes for this 300-million-acre area began a unique effort that sees the land the way Pluie might have. The Yellowstone to Yukon Conservation Initiative, launched in 1997, is a nonprofit that's been described as a network, a social movement, a mission, a vision. At two thousand miles long, Y2Y extends over five U.S. states and four Canadian provinces and territories. It comprises about seven hundred protected areas, anchored of course by Yellowstone but also including Banff, Canada's oldest national park, and Waterton-Glacier International Peace Park, on both sides of the border, the world's first international peace park, established at the urging of Rotarians from the United States and Canada. Y2Y now coordinates three-hundred-plus groups and individuals whose actions affect the land but who never worked together in any sustained way: Native American and First Nations communities, mining executives, foresters, ranchers, highway engineers, park officials, biologists, conservationists.

Y2Y is based on what animals like Pluie have shown us: that, as stand-alones and set-asides, even gargantuan national parks are too small to serve as containers for wide-ranging wildlife. Pluie—who kept moving in and out of protected areas—opened up another possibility, showing that even a divided landscape

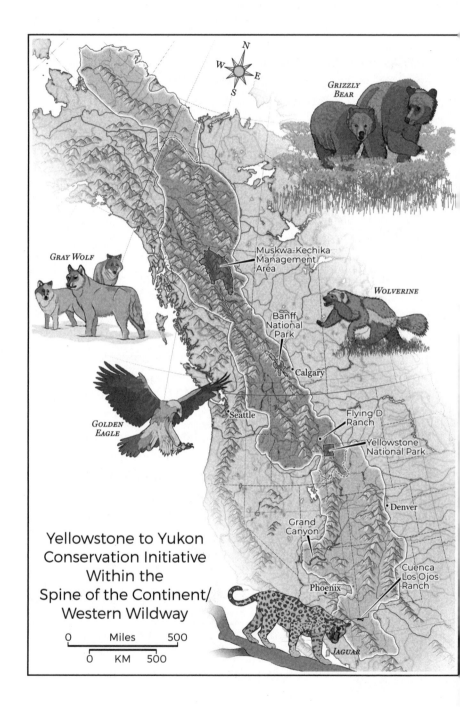

GRIZZLY
BEAR

GRAY WOLF

WOLVERINE

Muskwa-Kechika
Management
Area

Banff
National
Park

Calgary

GOLDEN
EAGLE

Seattle

Flying D
Ranch

Yellowstone
National Park

Denver

Yellowstone to Yukon
Conservation Initiative
Within the
Spine of the Continent/
Western Wildway

Grand
Canyon

Cuenca
Los Ojos
Ranch

Phoenix

JAGUAR

0   Miles   500

0   KM   500

can retain its wholeness if enough "movement zones," as Locke thought of them, are still in place, tying the protected areas together.

It took a century to understand this, that even half a landscape can reunite the natural workings of a region. Here was a Pluie-led way to move past what for so long had seemed inescapable, the displacement of a place, and instead find room for a new West that includes human and animal communities that "Progress" had dislodged. This was an end run around inevitability.

The first step came in December 1993, when Locke co-convened a small session of Canadian and American activists and conservation biologists, including Michael Soulé and Reed Noss (their cores and corridors looked at the land in a similar way—same focus, different lens). The next spring, the Y2Y piece Locke wrote for that meeting appeared in both *Borealis* and *Wild Earth* magazines. After that, Locke told Nick Walker, in *Canadian Geographic,* "it was like that high school science experiment where you drop a string into a solution and everything crystalizes on it right away." Since 1993, the amount of land within Y2Y that protects animals on the move has jumped from 11 percent to 21 percent, with improved conservation on another 30 percent. In Y2Y's office just south of Banff, Pluie's collar occupies an honored place.

The logical next step was even more of a leap: to scour the rest of North America for other Y2Ys. A tricky business, it turns out, identifying Pluie-sized places with real pull, magnetic places that can speak to the neighbors and unify antagonistic groups. Work at this scale started slowly, but lately it's been accelerating. There are now several names for it (all of them decidedly academic): large landscape conservation, landscape-level conservation, and landscape-scale conservation.

According to Jodi Hilty, a conservation biologist who is president of Y2Y, "Over the last decade, working at a 'large landscape scale' has become de rigueur in the arena of conservation biology and conservation implementation." There's also the Network for Landscape Conservation—more than one hundred and fifty groups and agencies and three thousand individuals working on over one hundred projects that think beyond boundaries.

In 2014, I was with more than three hundred presenters packed into a Washington, D.C., conference center for a noisy two-day National Workshop on Large Landscape Conservation, the first event of its kind, to hear about projects like Two Countries, One Forest (its own catchy acronym is 2C1Forest), dedicated to protecting an 82-million-acre forest that spreads out over four Canadian provinces and four U.S. states. I stood at the back of an overflow crowd when one of the 2C1Forest leaders said they were committed to extending Y2Y's work: "We are the very young East Coast cousin to Y2Y that one day will be its equal."

"Nature needs half," Harvey Locke explains in "Harvey Wants Half," Walker's *Canadian Geographic* piece, and Locke is cofounder of the Nature Needs Half movement, whose mission, Walker writes, is to "transform the relationship between society and nature and protect 50 percent of the planet by 2050." Which "is simply Y2Y gone global."

In 2017, several Canadian wildlife ecologists came up with a manifesto for what could be called "tall landscape conservation"— which is more like Y2Y gone 3D. The word they're using is "aeroconservation," referring to the ten miles above our heads, which they call the final ecological frontier. Every year up to a billion birds are killed in the United States by colliding with tall buildings, and as many as 50 million die from "towerkill"—flying into radio towers. The idea, according to an article by Lesley Evans Ogden in *New Scientist:* "to reclaim the air for its inhabitants, creating protected areas that extend into the sky and designing buildings to avoid death."

The wildest, greatest success story

Once upon a time wolves were more widespread than any mammal on the planet, except humanity. In 1600, shortly after the first European settlement in what's now the United States, there were something like 2 million wolves in North America. As settlement moved westward, no other animal was so relentlessly and successfully trapped, shot, and poisoned, usually with strychnine, which produces severe muscle spasms throughout the body. They were hunted with dogs, set on fire, and dug out from dens with what environmental historian Jon T. Coleman, in his book *Vicious: Wolves and Men in America*, calls "gruesome expressions of revenge, anger and dominion." He attributes this to the power of folklore—"wolf hatred" imported from Europe—and of newly acquired property rights. Wolves were killing the livestock that farmers and ranchers brought in to replace the great herds of buffalo, also eradicated, which wolves had always eaten.

Diane Boyd and a sedated wolf

Coleman mentions a day in 1814 when the great naturalist John James Audubon, then in his twenties, watched an Indiana farmer mutilate three wolves before his dogs ripped them apart. Audubon described it as "sport." Several years later Ohio proclaimed a war of extermination on wolves and bears. Bounty programs that allowed "wolfers" to kill one hundred thousand wolves a year during the 1870s were still in place and offering twenty to fifty dollars per wolf until 1965. By 1973, when wolves became one of the first species protected by the new Endangered Species Act, there were fewer than eight hundred in the Lower 48, all of them in the Upper Midwest, just below Canada.

Today the Upper Rocky Mountains, meaning Montana and Idaho, are considered a wolf-rich area, and gray wolves are spreading west and then south into Washington, Oregon, and California. Diane Boyd, a biologist whom Montana's *Flathead Beacon* called "the Jane Goodall of wolves," was the country's first female wolf trapper and tracker. She told the Wildlife Society that wolf recovery is "the wildest, greatest success story of any endangered species I can think of."

Boyd can be confidently categorical because the reappearance of wolves in the West came two decades after radio collaring and, more generally, the field of biotelemetry (detecting an animal at a distance). "It was an amazing time to be a researcher," Boyd told me. All together the University of Montana's Wolf Ecology Project, which Boyd was part of when working out of a remote cabin that "barely had walls," collared 1,681 wolves in a 92-million-acre area—with Boyd herself trapping "north of a hundred."

When Boyd got to Montana in 1979, there was only one collared wolf to follow, a female named Kishinena, who Boyd said had "walked down" from Canada, where there were tens of thousands of wolves; this was years before Canadian wolves were deliberately returned to Yellowstone, which, under a so-called predator control program, shot its last wolf in 1926. In the early 1980s, eight recurrent tracks in the Montana snow showed that Kishinena, whose collar had gone silent, had by herself (she had lost her mate) raised seven pups without losing one; they were dubbed the Magic Pack.

Basically all it takes for wolves to do well is for people not to kill them, or to kill fewer of them. Each wolf pack, wrote L. David Mech and Luigi Boitani in *Wolves: Behavior, Ecology, and Conservation,* can be considered a "dispersal pump that converts prey into young wolves and spews them far and wide over the landscape." With very few natural boundaries, Boyd notes in a study she coauthored, "only extremely large bodies of water or deserts appear to prevent wolf dispersal."

Most wolves disperse, not just Pluie, and they head off in all directions. It's this phenomenal ability, an innate, omnidirectional behavior, that helps explain their resilience even after getting "delisted" as an endangered species so that hunting can resume. Quite a few wolves travel Pluie distances, it turns out. A Yellowstone wolf made it all the way down to the Grand Canyon and most of the way back, covering 650 miles. This wolf, and others, have outlined a landscape down the full length of the Rockies from Alaska to Mexico, a distance called the Carnivore Way—a term invented by ecologist Cristina Eisenberg.

But Wolf Ecology Project data—a record of straight-line, end-to-end distances—may not tell the full story. Look again at the particulars of these journeys and more intricate patterns emerge. A Montana yearling also wandered south, drifting east to Colorado, where he was shot. As recorded, this was a shorter, four-hundred-mile trip, but to say that is inadequate. With his GPS monitoring factored in, the straight line of his journey bends left and right and takes on the shape of a sinuous, circuitous three-thousand-mile-long path. Some of the many routes wolves follow overlap, becoming so broad and taking in so much territory that Boyd thinks "corridors" is the wrong word for describing them. To me they sounded more like "swaths."

"When the blackbird flew out of sight," wrote the poet Wallace Stevens, "it marked the edge of one of many circles"—a stanza that is this book's epigraph and appears in *Where the Animals Go,* an award-winning book of GPS animal-tracking maps by Oliver Uberti, a cartographer and designer, and James Cheshire, a geographer. A lasting impact of telemetry is that now nothing can be thought of as entirely out of sight or beyond reach. "When you look at these tracks," Uberti told Marina Koren in *The Atlantic,* "you see in each individual path animals making decisions that are unique to them—what they like to eat, where they like to go." Which means these newer studies are less about what the observer is going through and more about what the animal is experiencing, the animal as storyteller.

Boyd told me she sees technology as two-edged. On the one hand, it's carried us closer to wolves and how they operate—"My respect for them grows and grows," she said. "They travel and hunt in packs because they don't have the massive strength of a grizzly bear, which can kill with a swat, or the precision bite of a mountain lion, which can crush a windpipe." Boyd, who loves dogs and raised pointers, told Perri Knize of *Sports Illustrated* in 1993 that if the canine world were a school, wolves would be taping notes to dogs' backs that say NIP ME. "Dogs are brain-dead wolves," she said at the time. "What if," she asked me, "every time you wanted to eat, you had to attack a flailing hoofed animal bigger than you, biting it repeatedly? And between meals had to contend with people and disease?"

On the other hand, reading about animal locations on a computer screen pulls us farther away from them. "You've got to go out and get dirty, and tired," Boyd pointed out. "Otherwise how will you ever know what a landscape is made of or find out what it smells like? In the Northwest at the edge of spring, just before leaves emerge, the buds of black cottonwoods are covered with an orange, sticky sap. This only lasts a week or two, but the air is filled with a sweet fragrance, not like a flower, not like a pine tree, but unmistakable." Some people think it smells like honey or apples or vanilla.

A landscape of opportunity

I met John Laundré, a wildlife biologist, during a 2017 conservation conference down a gravel road high up in the North Carolina Appalachians. Fall colors were just beginning to touch the peaks, visible even on a foggy day as we chatted about the other side of the continent. "Yellowstone, in 1996, was like two different countries—one at peace, the other at war," Laundré said. "The southern end of the park was like Disneyland, the elk bouncing along happily. In the north, the elk kept lifting their heads to look around, snatching bites of grass furtively, their calves close, huddled by their sides." That was the summer after wolves were reintroduced to the north part of the park; they hadn't yet made their way south.

Laundré was primarily a cougar biologist and just happened to be there, which opened a back door into wolf research. He'd gotten a doctorate studying coyotes in Idaho, and then, he says, "switched to cats," but received a small grant to follow the Yellowstone wolves. Laundré found they were doing far more than eating elk: they were imposing a novel and unsuspected pattern on the landscape, one that offered a rare glimpse into the inner lives of their prey. "Ecologists," he said, "don't like to incorporate behavior into their work. But this was inescapable, the awareness

Wolf pack and a bull elk in Yellowstone

and intelligence blatantly on view." Laundré coined a term—the "landscape of fear."

There's another way to understand how wolves changed Yellowstone—the so-called trophic cascades model, also known as the green-world hypothesis. "Trophic" comes from the Greek word for nourishment, and in ecosystems it refers to who's eating what. Trophic levels are a way of organizing the food chain, generally with plants on the first level, plant eaters on the second, and meat eaters on the third. The idea behind trophic cascades is that if a "who" is a keystone species, meaning one with an outsized influence, that "who" can cause a spillover effect on several different "whats." Specifically, predators are the ultimate protectors of plants—hence the name "green-world hypothesis"—because they keep in check the number of herbivores, the predators of vegetation.

As Laundré sees it, large carnivores like wolves, mountain lions, and grizzly bears are "tenders of the garden, the shepherds of the wild flocks," as he says in his book *Phantoms of the Prairie*. Case in point: elk, which are voracious consumers of the green world. They are grazers, animals that eat what's growing on the ground, and also browsers, animals that eat upward, meaning leaves and shoots and tree saplings as tall as they are—and a bull elk is five feet high at the shoulders.

In trophic cascade terms, it was predictable that elk numbers would fall as wolves came back to Yellowstone—and they did. But the drop was much greater than expected, given that wolves only kill about a quarter of the elk they attack. Elk that survived began eating less and having fewer calves, and this, Laundré realized, had something to do with *where* they were eating. Wolves hunt by chasing down prey, unlike mountain lions, which rely on stealth, stalking, and ambush. Elk began avoiding open grassy areas where wolves had a clear line of sight, even though that's where eating was easiest and food more plentiful. Instead elk withdrew to the forest edges, where pursuit was trickier. It was as if the geography of their lives had shifted.

"You've got to look at landscapes through the eyes of animals," Laundré said with emphasis. For seventy years the park without wolves was damaged and incomplete, but for the elk it was paradise.

Ungrazed after being abandoned by the elk, Yellowstone meadows began to regenerate. Yellowstone wolf biologist Doug Smith, who came to the park in 1994, a year before the wolves, remembers how willow and aspen shoots only reached his knees, and "now I can't see through it, it's like a forest," he told Katharine Lackey in *USA Today.*

This helped revive another keystone species: the beaver. Beaver dams create wetlands that improve water quality (wetlands are sometimes called "earth's kidneys") and provide a home for many plants, birds, fish, and frogs. Before the wolves, beaver had all but disappeared from Yellowstone because they need willow logs to build their lodges, and elk grazing and browsing prevented willow trees from growing high enough, thick enough. The elk's landscape of fear was for beavers a landscape of opportunity.

The trees' regrowth has been patchy. A study by Michel Kohl and Dan MacNulty, Utah State University ecologists, suggests that even the landscape of fear has its comings and goings, ups and downs. Elk can sense a lull in wolf activity in the middle of the night, making it possible to steal out into open areas for a few hours. This, researchers say, can temporarily flatten the landscape of fear. It will take time—decades, some biologists think— before the full workings of Yellowstone are understood.

In the meantime what's exciting to Doug Smith, as reported by Brodie Farquhar in *National Park Trips* magazine, is the "rare, almost unique opportunity to document what happens when an ecosystem becomes whole again." Smith told Farquhar, "In the entire scientific literature, there are only five or six comparable circumstances." What's exciting for Laundré is Yellowstone's demonstration that, as he told me, "fear is one of the strongest forces in evolution, something that arose as soon as one thing started eating another. Fear forces animals to use habitats that are subprime, strengthening their adaptability. The expulsion from Eden has been one of life's greatest gifts, a constant pressure—you can starve for a day and survive, but relax your vigilance at the wrong moment and you are no more. Fear drew some animals, like our primate ancestors, up trees for greater safety. Other animals took to the air on wings."

Cristina Eisenberg, in her book *The Wolf's Tooth: Keystone Predators, Trophic Cascades, and Biodiversity,* sensed a foreshadowing of this in lines from a 1940 poem by Robinson Jeffers, and it became the epigraph for her book: *"What but the wolf's tooth whittled so fine / The fleet limbs of the antelope? / What but fear winged the birds, and hunger / Jeweled with such eyes the great goshawk's head?"*

The most adventuresome

"Carrying pepper spray has already become indispensable for hikers, hunters, and others in many parts of Montana, Idaho, and Wyoming," *The New York Times* reported in 2017, in a piece about how long-isolated grizzly bear populations in Yellowstone and Glacier National Park are expanding, and without any human intervention will likely join up sometime in the 2020s. "Practice!" warns a bear-safety page on the Yellowstone website about encounters with these eight-hundred-pound animals, which can run forty miles an hour. "Use an inert can of bear spray to practice removing it from your holster, removing the safety tab with your thumb, and firing." A grizzly's favorite day of the week is Chewsday, according to Richard H. Yahner in his book, *Fascinating Mammals.*

Grizzlies were declared a threatened species (approaching the brink of extinction) in 1975, two years after wolves became an endangered species (at the brink). Back then, there were 136 grizzlies in Yellowstone and now at least 700 roam the park, maybe 1,000. Yellowstone and Glacier are about 375 miles distant, but by 2017 the most adventuresome bears from either park were only 50 miles apart.

As gray wolves, mountain lions, grizzly bears, and other large roaming predators return to landscapes throughout North America, where a hundred years ago they were considered varmints (the word originally meant "worms"), can people adapt to a shift in their own geography, a de-Edenizing? I put this question to Diane Boyd. "Well, I'm down in the trenches with ranchers who hate wolves," she said. "Some say wolves shouldn't be here and are only present because that's where the government put them. All I can do is offer good science."

The resistance she faces: "We don't need wolves in our backyard and grizzlies in our front yard," one rancher said at a hearing. "Don't hand over purple mountains' majesty and fruited plains to environmentalists who put biota and bugs above humans." One anti-grizzly article called them "a Eurasian species that might not even belong here."

This is life at the Half Earth interface. "One of the greatest things about our country is that people can do what they want," Boyd told me. "But it's a trade-off, and I had to learn this the hard way. Thirty years ago I had a dog who would not stay home. A mountain lion got her. Now I'd fit any dog with an e-collar, which administers a mild buzz if she wanders too far away." As for bears, she says, "You can't have bird feeders out in the summer, when bears are active. You don't need chicken coops or beehives, which are attractants, billboards for bears. Buy eggs and honey! Make your living space unalluring for bears! You can work with some residents—these are not big changes. But there are always new people coming in. The turnover is constant. Is it possible to keep up with it?"

Migrations made visible

It was so makeshift you could hardly call it a technological break-through, but one of the first attempts to be precise about the patterns of bird migrations had to have been the work of John James Audubon. In 1803, while still a teenager in Philadelphia, he tied "light silver thread" to the legs of eastern phoebes about to leave their nests. A decade later Audubon was indifferent to wolf butchery, but here he made sure the thread was "loose enough not to hurt." The following year, at the same nests: "I had the pleasure of finding that two of them had the little ring on the leg."

Audubon's eastern phoebes

He had no idea where these plump, gray songbirds had been in the meantime—we now know they winter in the southern United States and northeastern Mexico—but he validated the principle of "philopatry," meaning the tendency to remain in or return to familiar ground. Julius Caesar used the same philopatric principle nearly two thousand years earlier to send messages by homing pigeons during his conquest of Gaul. In the sixteenth century, royal falcons wore silver bands as a badge of honor—so that when a silver-ringed peregrine turned up one day on the island of Malta in the Mediterranean Sea south of Sicily, it was clearly a bird that belonged to King Henry IV of France. Defying philopatry, the peregrine had escaped the day before from Fontainebleau near Paris, 1,050 miles to the northwest.

Systematic bird banding dates to the end of the nineteenth century. In North America it took on a continental focus because of the Canadian-American Migratory Bird Treaty Act of 1918, to prevent "indiscriminate slaughter" after an outraged public response to a lethal fad—dazzling white curving plumes were commanding huge prices as adornments to women's hats. This nearly exterminated birds like the snowy egret, the prowler of tidal wetlands and considered the most elegant of herons.

The treaty and follow-up legislation, which make it "unlawful to pursue, hunt, take, capture, kill, possess, sell, purchase, barter or transport any migratory bird, or any part, nest or egg of any such bird," are credited with saving millions, perhaps billions of birds. And snowy egrets, also known for their black legs and bright yellow feet, are no longer a rarity and are seen every day, for instance, by commuters taking trains between New York City and Newark, New Jersey.

In 1920, a U.S. government program run with Canadians coordinated the bird-banding efforts in both countries. Its self-effacing administrator, Frederick Lincoln, an ornithologist whose only hobby was his work, carried on his research so quietly, an obituary in *The Auk* said, few people realized how much he knew. In ten years he gained a Pluie-like insight into vast landscape and airscape patterns enveloping a territory larger than Y2Y. Patterns in this case demonstrated not by the meandering wanderings of an individual but by the steady travels of millions. "Recovery of banded ducks and geese accumulated so rapidly," Lincoln noted, "that by 1930 it was possible to map out the four waterfowl flyways' great geographical regions, each with breeding and wintering grounds connected by a complicated series of migration routes."

The four flyways span the continent, top to bottom—two hugging the coasts and two in the middle—all dating back to the beginning of the Ice Age 2.6 million years ago, when glaciers dominated the center of North America. Also apparent was that, while breeding and wintering areas might be separated by thousands of miles, north to south, there were, philopatrically, regular stopover points along the way—which provided a framework for setting up wildlife refuges en route and not only at either end. This was the birth of aeroconservation—and it all got done a generation before radio collaring, the result of hundreds and later thousands of bird banders who placed lightweight aluminum rings (nearly 5 million by 1945) on birds as carefully as Audubon had attached his silver thread to phoebes about a century and a half before.

Angels

In the early days of World War II, British use of radar (an acronym for radio detection and ranging, and then brand-new) gave England an edge over the Nazis, according to Britain's Imperial War Museums, providing early warning of German air raids, allowing Royal Air Force fighter planes time to get off the ground. Radar is a kind of echo locator—radio waves bounce back to an operator from any object they encounter.

But in 1940, during the Battle of Britain, there was a problem: the presence of extraneous phantom echoes variously referred to as ethereal, mysterious, and enigmatic, which seemed to be coming from otherwise undetectable objects on or over the English Channel. These signals could be moving at five or eighty miles an hour, and either with or against the wind. Some disappeared quickly, others persisted over a thirty-five-mile-long course, and according to an essay by Anthony D. Fox and Patrick D. L. Beasley in *Archives of Natural History*, "many would annoyingly wax and wane." The signals were not related to weather conditions. Could they be fast-moving ships? Technicians took to calling them "angels."

An English evolutionary biologist, David Lack, considered one of the greatest ornithologists of his time, happened to be part of a British Army research group in 1940. He was able to figure out that the "angels" were birds and even pointed out that the erratic nature of the signals might be related to wingbeats. A colleague of Lack's, George Varley, an entomologist who later was best man at his wedding, proved the point when, through a telescope, he saw a flight of gannets, the largest seabirds in the North Atlantic (white with yellow heads and six-foot wingspans) at the exact moment and location radar screens were picking up an "angel." Skeptics were unconvinced. The number of migrating birds flying at night was not yet well understood or studied. Pre-radar, the only way to detect nocturnal migrants was "moonwatching"— trying to count silhouettes and calculate the number of birds flying in front of a full moon.

These days, birds taking to the air half an hour after sunset, a preferred departure time for many migrants, show up spring and fall on NEXRAD's screens, which receive signals from the 160 Next Generation Weather Radar stations operated by the National Weather Service. A pulse of fifty thousand purple martins, the largest North American swallows, known for their speed and acrobatics in the air, may take the form of a spreading blue doughnut on the screens, according to a piece posted by the Cornell Lab of Ornithology's BirdCast project. Moth and ladybug migrations, and the fleeting flights of mayflies, which live only a day or two, are also made visible by NEXRAD.

Now that migrations have become visible, a next step in radar ornithology, pioneered in 2018 by the BirdCast project, is a national map showing three-day migration forecasts: "Brighter = more birds," with yellows and reds showing a greater concentration of activity. Birds like to fly in clear skies and bunch up on the ground when the weather turns bad, so the online BirdCast maps, the Cornell Lab reports, "compile more than 20 years of data on

when migrants are on the move and mash it up with meteoro-
logical data including wind speed, direction, barometric pressure
and temperature. You no longer have to have a Ph.D. in migration
ecology to use all these variables."

The wild heart of North America

"Let us not have puny thoughts," wildlife biologist Adolph Murie
wrote in 1965 about Denali National Park in Alaska, then known
as Mount McKinley National Park. "Let us think on a greater
scale. Let us not have those of the future decry our smallness of
concept and lack of foresight." Murie was part of a famous and
close-knit clan of naturalists. He and his half brother, Olaus
(known as the father of elk management), married half sisters.
Margaret (Mardy) Murie, Olaus's wife, helped establish the Arctic
National Wildlife Refuge (she has been called, somewhat oddly,
the grandmother of the conservation movement).

Adolph and Olaus Murie by Alfred Eisenstaedt, 1961

In the summer of 1939, a couple of decades before radio collars, Adolph Murie, equipped only with binoculars, notebooks, a telescope, a still camera, and a movie camera, hiked more than 1,700 miles through McKinley National Park studying wolves for the National Park Service. Dall sheep (large white sheep with elegant curling horns that live on the park's steep, windswept ridges) were in sharp decline, and wolves were getting blamed. For Murie this began a meticulous two-year study—he had a pair of skis for winter—work that allowed him to spend about 195 hours observing a single wolf den from half a mile away, including one nonstop vigil that lasted 33 hours.

Murie's 1944 book, *The Wolves of Mount McKinley*, the first in-depth examination of the species, is on sale to this day, considered a classic and a turning point. Murie concluded that late-winter weather had been killing the sheep, not wolves, and that wolves were an essential part of this ecosystem. The park eliminated its wolf-eradication program several years later, and this served as the precedent Yellowstone could follow when, half a century later, it too welcomed wolves. "It is true basic research," Olaus wrote about his brother's book in 1961. "It means living with the animals, trying to think as they do." On-the-ground contact with animals could not be hurried but had to be pursued, Olaus said, "for as long as it takes to get the desired information."

So revered is Adolph (Ade) Murie's approach that Sanctuary River Cabin No. 31, the small remote hut that was his headquarters, has been on the National Register of Historic Places since 1986. A Ken Burns documentary series, *The National Parks*, quotes writer Terry Tempest Williams, a Murie admirer:

> I often think if we were to send for a representative of our species to meet with the animals, we would send Ade because he's a man who knows how to listen. He was a man

who understood stillness. And more than anything, his curiosity and his extraordinary sense of science opened up the landscape in a new way for all of us. He saw the land as a set of relationships, nothing in isolation, everything connected.

Starting in 1959, Frank Jr. and John Craighead, handsome identical twins and biologists who became twentieth-century "eco-idols," spent twelve years studying grizzly bears in Yellowstone. They and their team hiked 162,000 miles—far more per year than Murie had. Like Murie with the wolves, their goal was a "rendezvous with grizzlies in their most intimate moments." Before setting out, they practiced pull-ups in case they had to shinny up a tree to get away from a bear. The big difference from Murie's research was that the Craigheads were following animals they themselves had tagged. In 1961, Frank and a couple of friends developed the first simple, workable radio collars for large animals. Previously, the typical mode of communication

The Craighead twins, Frank and John, 1952

between people and grizzlies, John Craighead wrote in *National Geographic*, was by way of a rifle bullet.

The twins were born in Washington, D.C., and after high school drove west. "The day our old '28 Chevy topped this hill in Wyoming and we spotted the Tetons," John Craighead recalled in Mike Lapinski's book, *Grizzlies and Grizzled Old Men*, "it was like our souls got sucked right into the Rocky Mountains." (John died in 2016 at the age of one hundred.) The brothers wound up living in identical, side-by-side log cabins they built outside Moose, Wyoming, near Grand Teton National Park, where the family, through the Craighead Institute, is still involved in wildlife research.

The twins had many interests: an early love of falconry led a maharajah's son to invite them to India when they were in their twenties; during their bear studies they published a field guide to wildflowers; and in their fifties they wrote much of the Wild and Scenic Rivers Act of 1968. Late in life, John told Vince Devlin, a reporter in Montana, that radio collaring probably boosted their data fiftyfold. "There was just no comparison to how much you could learn when you had the radio," he said.

The Craigheads' most lasting accomplishment might be their map-changing discovery based on what the bears revealed about the shape of the land. One male grizzly could range over 900,000 acres. After collaring 256 bears, it became plain that, for many Yellowstone grizzlies, the park and the adjacent Grand Teton National Park and the national forests that surrounded the two parks were all the same landscape. The Craigheads drew a map of the park and plotted a rough hexagon around it—places the bears had made their way to. In Frank Craighead's 1979 book, *Track of the Grizzly*, he introduced a term for this area: the Greater Yellowstone Ecosystem (GYE). Yellowstone has some 2 million acres; the bear-defined Greater Yellowstone Ecosystem (the name stuck) has 22 million. The Greater Yellowstone Coalition, set up in 1983, raises millions of dollars every year to protect what they call the wild heart of North America.

Right underneath our noses

In 2013, a National Geographic video about Greater Yellowstone wildlife went viral on YouTube, with close to 8 million views. The video wasn't about rewilding an ecosystem (adding back a missing species). You could say it was about un-de-wilding—preventing the loss of something that has never gone away, which is just as crucial to 50 by '50. The video focused on the second trophic level (plant eaters), less explored, less controversial—in this case, mule deer, one of Yellowstone's charismatic big game prey species, not the large predators that consume them. "Big game" is a hunting term; for African hunters it refers to the "Big Five," the hardest animals to hunt: lions, leopards, rhinos, Cape buffalos, and elephants.

The biggest game animals in North America are fast and elusive, always up on their toes (hooves are hardened, elongated nails)—moose, elk, white-tailed deer, and their larger, western cousins, mule deer. In 2011, Hall Sawyer, a research biologist, was asked to make a routine study of a herd of mule deer. These Wyoming mule deer were assumed to be a resident population, pretty much stationary, maybe moving around as much as forty miles but spending their whole lives within the Red Desert. Lightly populated with people and heavily populated with wildlife, the Red Desert is a vast, unfenced area of sagebrush and grasslands, buttes, and sand dunes in the southwest of the state. The Audubon Society considers the place "incredibly unique," and it was once proposed as the site for a million-acre national park.

A mule deer stotting

Like a gazelle, a big-eared mule deer can stot or pronk or prong—terms that describe the way they bounce along and spring into the air with all four feet off the ground. It's something they can do straight uphill to escape a predator. That much was already known about mule deer behavior. In January 2011, Sawyer put GPS collars on forty Red Desert mule deer, and in the spring sent up a pilot to pick up their signals and report on their whereabouts. But there were no whereabouts. Sawyer was at his desk in Laramie, Wyoming, when he got a text: the deer were missing.

Heading north for forty miles, the pilot only picked up a couple of signals. Sawyer wondered whether he'd used a bad batch of collars. The pilot was heading back to his base even farther north and, figuring it wouldn't hurt, Sawyer asked him to keep scanning. An hour later, the pilot called back—he found more of them, one hundred miles from "home." (Echoes of Pluie.)

The story, as it emerged over the next couple of years, was complicated. Some mule deer did stay put year-round; some went north to a mountain range where the pilot picked up the first signals. But most continued on to the Hoback Basin, part of a national forest south of Grand Teton, three thousand to four thousand feet higher and 150 miles away from where they'd started.

What Sawyer had discovered was the longest land animal migration in the Lower 48. "We kind of think we know what's going on," he says with a grin in the viral video, "and here we have hundreds of animals migrating a hundred and fifty miles across public and private lands right underneath our noses and we didn't even know about it." Not once but twice each year, it turns out, in a procession that could be six thousand years old. "That it still happens is remarkable," said Emilene Ostlind in *High Country News;* she is an environmental journalist who's been tracking the migration. "Around the world, long-distance animal migrations are disappearing as human development blocks and fragments migration corridors." And this particular migration is happening not in the Serengeti but right in the middle of North America.

Then came a mule deer who never got a name, unlike Pluie, and was simply called Deer 255. In 2016 she was collared as part of a follow-up study; she reached the Hoback but didn't stop there, continuing on through Grand Teton north and west to the outskirts of Island Park, Idaho, adding another 90 miles to her journey (and picking up a nickname, Island Park Girl). At that point her collar failed. Was this a fluke? Had she joined an Idaho herd? In March 2018 she was captured and collared again, back in the Red Desert on her way up to Island Park. In 2019, pregnant with twins, she set off yet again but this time stopped short of Idaho, after traveling 200 miles. The story remains incomplete. At least for Deer 255, the 150-mile-long migration was more like a 240-mile migration, and researchers suspect this could be true for hundreds more mule deer.

Clearly the Red Desert, thought to be well beyond the southern boundary of Greater Yellowstone, was for mule deer an integral part of the ecosystem. Matt Kauffman, the biologist who leads the Wyoming Migration Initiative, a group of researchers who disseminate data as rapidly as they collect it (they made the Nat Geo video), told the *Casper Star-Tribune* that it's amazing and exciting "there are still these discoveries to be made," patterns "we are struggling to document and understand."

Surfing the green wave

What's also clear is that movement matters, and migration is what keeps wildlife thriving in Wyoming's otherwise forbidding landscapes. High-elevation forests and meadows are lush in summer but get up to eighteen feet of snow in winter. At a lower elevation, sagebrush areas, like the Red Desert, get only thin snow that keeps shrubs exposed and browsable, but in summer, temperatures can reach one hundred degrees. The first trophic level (plant life) is the only habitat connector, specifically the luxuriant green wave of spring growth. Like an annual tide, it can be tracked across the continent by satellite mapping, and in Wyoming the green wave always flows uphill as snows melt.

Plants are what make the Wyoming year unfold—they guide, lure, and beckon the wild flocks. Superimposing green-wave satellite images on the mule deer migration route, it became easy to see just how synchronized the two are. Researchers call it "surfing the green wave," with mule deer reaching new plant growth right as the plants they most favor, poa grass and sticky purple geranium, are an inch high. It's the exact moment these plants are most succulent and high in nutrition but still low in harder-to-digest fiber. Each spring and fall, the mule deer wind up spending two months or longer on the move, always with stopovers, along a route their speed would otherwise allow them to dash through in two or three days.

Greater Yellowstone moose, bighorn sheep, bison, and elk are fellow surfers of the green wave. It looks as though elk may lag behind the peak of it; they're big animals and can take bigger bites and handle roughage more easily. Bison, instead of following the green wave, graze and trample the spring grass in groups of hundreds or thousands, which regenerates the grass and creates what Mark Hebblewhite, a University of Montana ecologist, calls "a state of perpetual spring."

Back in the 1990s, Sawyer was able to trace the 120-mile-long Path of the Pronghorn, a route used by an antelope-like species ancient enough to be able to run sixty-one miles an hour for reasons that no longer apply. (Some researchers think they had to be faster than the American cheetah, which went extinct twelve thousand years ago.) It was a time when oil and gas development was already on the rise in Wyoming, and, Sawyer told *bioGraphic* magazine, "I was just trying to put myself in a pronghorn's hooves." But these pronghorn summer in Grand Teton, a secure landscape, and in 2008 the U.S. Forest Service designated the forty-five miles of the path that go through a national forest as the first-ever federally protected national migration corridor. The hope is that it will stay permanently undeveloped and unmined and undrilled.

But the mule deer Sawyer collared in 2011 winter and summer on land not part of any national park or protected wilderness, and their path, though far older than any ownership, crosses federal- and state-owned land as well as private property in the hands of forty-one separate proprietors, according to Kristen A.

Schmitt in *Smithsonian* magazine. The mule deer also have to get past nearly one hundred fences and on several highways need to outrun traffic as fast as they are (not always successfully). Greater Yellowstone migrations—and there are dozens, including a "superherd" of elk made up of nine subherds—have been described as the pulse of the park and, according to ecologist Arthur Middleton, are "the greatest wonder of the Yellowstone ecosystem that people haven't seen, until now."

Right-of-way

Among hikers, trail etiquette dictates that those moving downhill step aside to allow uphill hikers the right-of-way; it's easier for the downhill hiker to break stride. Good trail manners, such as this, date back at least a century. But what happens when one species—the human newcomer, say—meets another on pathways the newcomers never knew existed? First came an awareness of swaths, wide tracts used by dispersing, roaming, returning predators. Then came the discovery of paths, narrow bands of land used by prey (migrating mammals moving back and forth between places in search of food). Swaths and paths weave together in intricate patterns.

What's at stake, it turns out, is the persistence of persistence. For instance, there's the Red Desert and the path the mule deer follow. It could be overwhelmed by a gray wave of development—houses, oil wells, new roads—severed in one or more spots by people and groups and agencies not in the habit of noticing one another or the animals moving among them. Already it's like the mule deer have to navigate a Jenga tower that multiple players could collapse at any time.

In the 1890s, George Shiras III, son of a Supreme Court justice and himself later a congressman, set up a trip wire attached to a camera and a dish of magnesium flash powder in the Pennsylvania countryside at night. When a passing deer activated the camera, the powder fizzed and then exploded, and Shiras got a picture of a leaping animal. He called this makeshift outdoor chem lab "camera hunting" and the explosions a "glowing moon."

Shiras had invented the camera trap, which gave people the ability to sneak up on something without being there and capture a close-up image of an animal too far away or too hidden to see. In the 1920s, using flashbulbs instead of magnesium, Frank Chapman from the American Museum of Natural History inventoried animals on an island in Panama with a camera trap rather than a gun. "A census of the living," Chapman said, "not a record of the dead."

These days, camera traps are motion activated, using infrared sensors that don't startle the animals. The Wyoming Migration Initiative has teamed up with National Geographic photojournalist Joe Riis, and his high-def videos, using camera traps, are stunning, eye-level, David Attenborough–worthy portraits of the hard-won treks the mule deer, elk, and pronghorn take. Riis shows the animals struggling through winter snows; battling fast-moving, snowmelt-laden spring rivers; plodding uphill with tongues hanging out; and covered with dust in early heat. It was Riis's Greater Yellowstone video that went viral.

More is lost than a place or an animal population if a migration path is shattered. Greater Yellowstone migrants move through a landscape of living, breathing memory. Green-wave surfing is not instinctive, but an acquired skill, one that takes generations to develop. As Ed Yong, *The Atlantic*'s science writer, wrote in a 2018 article, "Humans Are Destroying Animals' Ancestral Knowledge," "Individuals learn to move through the world by following their mothers, and then augment that inherited know-how with their own experiences."

This became known when bighorn sheep, whose pair of massive curling horns weighs up to thirty pounds, and who are so sure-footed they can stand on a ledge two inches wide, were moved back into areas they'd been hunted out of and then radio-collared. Yong cites a 2018 *Science* magazine study showing that less than a tenth of the new sheep migrated, while a quarter of the herds in place for thirty years did. The only expert surfers were the very long-standing herds, those that had been part of the landscape for centuries.

Ecologist Brett Jesmer told Yong that for sheep, full working knowledge of a new environment needs fifty to sixty years. It also seems to be a matter of learning to anticipate—knowing that beyond a risky patch is the excellent forage they remember. Moose, who weigh up to nine times as much as bighorn sheep, for some reason take longer to master new territory, as long as a century. Greater Yellowstone is more than a natural landscape—for the animals, it's a cultural one. As ecologist Carl Safina wrote in *The Guardian* about sperm whales, macaws, and chimpanzees, animals "are born to be wild. But *becoming* wild requires an education."

The internet of animals

Martin Wikelski, an ebullient German ornithologist, has spent two decades working on a multimillion-dollar global tracking project, ICARUS (International Cooperation for Animal Research Using Space). It's an acronym Wikelski chose deliberately after NASA wouldn't be part of it—was he flying too high, taking on too much? Wikelski was sure ICARUS would be up and running years ago, but then came huge setbacks, then getting somewhere, then falling short, then scrambling ahead—all while feeling, he told me, in the middle of something "furtive yet tremendous." Throughout the tracking and telemetry era, animals have been telling us things, but they've never had anything like an equal voice in the conversation. Wikelski was putting together what he called the internet of animals, something with a simple, sweeping purpose: anytime people accessed it, they would be learning from animals, alongside animals.

Wikelski formed a partnership with Roscosmos, the Russian space agency, and the German Aerospace Center, and he is now director of the Max Planck Institute of Animal Behavior, near Lake Constance, where Germany, Switzerland, and Austria converge. He has a staff of five scientists, a slew of postdocs and grad students, and hundreds of volunteers.

If all goes well, Wikelski says, in 2030 ICARUS will upload information every day from hundreds of thousands of animals around the globe. That way, wild animals would become our companions—"Humanity as a whole is thus being provided with a Seeing-Eye dog," he spells out in a foreword to *Animal Internet: Nature and the Digital Revolution,* a book by German writer Alexander Pschera. "There is no better information system than that of animals, informed as it is by the unparalleled diversity of their sixth sense." What animals have always known "will become an inextricable part of the common cultural framework of humankind, right alongside libraries and museums and the Internet itself."

Extending our awareness through animals—there's an example from 2,400 years ago Wikelski likes to cite in talks. The Roman republic was saved from invading Gauls who, on a moonlit night, scaled an unguarded cliff at the back of the Capitoline Hill where many Romans had taken refuge. Sleeping soldiers and even their watchdogs heard nothing, but the Capitoline geese, a sacred flock on the grounds of the Temple of Juno, honked and squawked so loudly they woke one soldier, Marcus Manlius, whose Roman family name is derived from the Latin for "morning." Manlius rushed to repel the attack just as the first Gauls were reaching the top of the cliff.

Here's a more well-known example: canaries. In 1913, at the suggestion of John Scott Haldane, a Scottish doctor, British coal miners began taking canaries into the mines as sentinels. Canaries are famously far more sensitive than people to colorless, odorless, poisonous carbon monoxide. As Kat Eschner points out on the *Smithsonian* magazine website, canaries need so much oxygen to fly, their lungs are absorbing air even when they exhale.

Wikelski says the inspiration for the internet of animals came to him at the University of Illinois at Urbana-Champaign, after conversations with a retired electrical engineer, Bill Cochran, and a retired radio astronomer, George Swenson Jr. Wikelski had been told particularly to steer clear of the engineer. "If Cochran shows up, run," he was told. "That old dude is evil." But when Wikelski, then a young assistant professor, met Cochran and Swenson in 1999, he told me he found both to be "the most solid people I ever met, with a perspective that was thirty to fifty years ahead of anyone else. Bill was a visionary, someone who thinks globally. One of the most fertile minds I've known in my life."

In 1959 Bill Cochran devised a radio collar a rabbit could wear. By the time he heard about a biologist who'd been awarded a hundred-thousand-dollar grant to come up with a rabbit collar, Cochran, on a fifteen-dollar budget, had already built one for eight dollars. Then, in 1962, he attended the first international biotelemetry conference, held at the American Museum of Natural History in New York. In the middle of it, during a casual discussion over drinks one night at a bar on the Upper West Side,

Cochran came up with the idea for satellite telemetry, something that wouldn't exist for another sixteen years.

In the 1950s, the Soviet Union launched the first artificial satellites, the beachball-sized *Sputnik,* and the slightly larger *Sputnik 2,* which put a two-year-old dog named Laika (loose translation: "Barker") into orbit. The high-pitched *beep-beep-beep* radio signals sent back to earth by the satellites were heard around the world. (Sadly, Laika, said to be "quiet and charming," died from overheating after a few hours aloft.)

"The reverse is true, too," Cochran said at the bar—signals from the ground could be picked up by satellites, and wouldn't that make animal telemetry a whole lot easier? Animals with transmitter collars, like his rabbit, were also sending out radio signals, but you had to be nearby to pick up the signals on a receiver. But a satellite, sweeping around the globe and passing over territory bigger than any animal path, could receive these ground-based signals from far away.

"I'm just an engineer," Cochran told me in 2019. "If I didn't

Bill Cochran's rabbit harness

think of the idea, some other guy would have." After the 1962 conference, Cochran didn't write up anything about satellite telemetry, but his ornithologist boss did, suggesting that the ideal prototype would be a study of a Laysan albatross, a little-known soaring Pacific Ocean bird with a six-foot wingspan. Cochran devised a four-ounce harness for this ten-pound bird—the rule of thumb is that any tag should be no more than 3 percent of an animal's weight, so as not to interfere with its movement. Cochran also arranged with a contact at Lockheed to piggyback this experiment onto a satellite they were already planning to launch. The total proposed budget for the albatross project was $25,000, but when it came up for review by the National Science Foundation, it was the last item on the agenda at the end of a long day and was treated as comic relief. Everyone had a laugh, voted it down, and went home.

A couple of years later, in 1964, when the Craighead twins were out in Yellowstone attaching bulky collars to grizzlies, Cochran, working with another ornithologist, put together a radio tag small enough and weighing little enough (seven-hundredths of an ounce, less than the weight of a dime) to attach to a Swainson's thrush. This is a shy forest migrant with a fluting, whirling song that seems to bounce from tree to tree, sometimes sung so softly it sounds like the bird is much farther away. For Cochran, this began forty-two years of tracking migrating birds over a hundred thousand miles, following birds that can fly twenty miles an hour (fifty with a tailwind).

He attached miniature radio tags to the birds' backs using false-eyelash adhesive, which allowed the tags to fall off after ten days and let the birds continue on their way unencumbered. "This was a major issue for me," he said. "I always felt guilty about interfering with their lives." Sometimes Cochran flew in a small plane. Mostly he drove around at night with a direction-finding antenna on a pipe sticking through the roof, turned by a handle from inside the car. Usually a student volunteer did the driving while Cochran worked the antenna, but if he was alone he taped a road map to the steering wheel and adjusted the antenna with his right arm.

One time he and a volunteer drove 940 miles over seven straight nights after a Swainson's thrush that finally disappeared into Manitoba, Canada, at a point where there were no roads across the border. Cochran has never had an accident. He's been mistaken for a drug runner and once flew through a thunderstorm so intense it convinced him to radio in his will. Now in his late eighties, Cochran is still writing up his results.

In 1999 Cochran took Wikelski on one of his all-night Swainson's thrush runs. When they got to Cochran's house the following afternoon, Wikelski found himself remarking that everything would be so much simpler if it could all be done remotely from space.

"Oh," said Cochran, "we had a plan for that thirty years ago."

As he listened, Wikelski was particularly impressed that Cochran kept working on making tags smaller and lighter, because most birds and insects and, in general, most animals, are small and would stay invisible unless they too could be tagged. "But there wasn't any brilliance involved," Cochran said to me. "Tinier parts kept coming on the market, from engineers designing for other engineers. Like anyone with any empathy for animals, we were always shooting to improve their comfort." Wikelski told Jesse Greenspan in *Audubon* magazine that it was amazing to think how much money Cochran could have made from his transmitters, but instead he gave them away "as a present to humanity."

George Swenson, the retired astronomer, introduced Wikelski to radio astronomy. This discipline, unknown until the 1930s, is all about making tangible the invisible sky—listening for and mapping faint radio signals from celestial objects, signals that cannot be obscured by daylight or interrupted by nearby clouds or faraway cosmic dust (things that can happen with optical-light telescopes). Radio telescopes can tune into phenomena that don't even give off visible light, like the large clouds of hydrogen gas that serve as the birthplace of stars.

For six decades Swenson flew around the continent, squeezing himself into a small private plane (he was six foot four). He built a four-hundred-foot-long radio telescope for the University of Illinois that found two supernova remnants, and he cataloged more than one thousand radio sources from beyond the Milky Way.

On a leave of absence in the 1960s, Swenson chaired the design team for the National Radio Astronomy Observatory's Very Large Array (VLA), a deep-space listening device. The VLA, built in a New Mexico desert a mile and a third high, is perhaps the most productive radio telescope in the world. Completed in 1981, upgraded in 2012, it consists of twenty-seven antennas plus a spare, each eighty-two feet across, to form a compound eye or ear on the universe. It has detected ice on Mercury, which has a night temperature as low as −279 degrees Fahrenheit, and pointed to the possibility of a black hole hidden in the center of the Milky Way, 2.6 million times more massive than our sun.

What Swenson did for Wikelski was give him that final push to start the animal internet. Keep in mind, Swenson told Wikelski, that very small sounds and very faint signals can only be detected by very big ears—and then go ahead and turn radio astronomy upside down. Listen to the earth, not the sky.

Swenson and Wikelski (with four coauthors) published the idea in 2007 in the *Journal of Experimental Biology:* "Wildlife biologists could point antennas towards earth from near-earth orbit to locate small radio transmitters attached to animals." By then, Cochran's ever-smaller eyelash-adhesive radio tags were no bigger than a thumbnail; the animal internet's solar-powered tags will be even smaller.

In their article, Swenson and Wikelski wrote, "Given the potential for this empowering technology to transform our understanding of the natural world, we have formed a global initiative to support its deployment: the International Cooperation for Animal Research Using Space." (This was the first mention of ICARUS in print.) In private talks with Wikelski, Swenson was blunter, saying, "Why are you guys in ecology so stupid that you won't get together as a global community and do for the planet what we radio astronomers have done for the universe?"

In August 2018 two Russian cosmonauts took an eight-hour, thirteen-minute space walk outside the International Space Station, 240 miles above the earth, and unfolded ICARUS antennas—two are six and a half feet long—capable of receiving bursts of

Martin Wikelski tags a scarlet macaw.

information from more than 15 million tagged animals. Plus another, shorter antenna, a transmitter that can tell tags anywhere how to calculate what's called the "contact window," the next time the space station will soar directly over their area.

There's also the sheer volume of what is being said by each of the tags, all talking at once like loud conversations in a crowded bar. Barely adding to their weight, tags can be outfitted with something like a Very Large Array of instruments—miniaturizations of centuries-old inventions such as accelerometers, magnetometers, gyroscopes, and pitot tubes, which measure air speed. Biologging, this process is called. "We can now document birds' flight behavior," says a *Science* article written by Wikelski and three coauthors, "as if they were airplanes carrying advanced aerospace technology." They add, "A golden age of animal tracking science has begun."

What more can animals tell us?

Aristotle thought that European redstarts transform into European robins (there is some resemblance; both are small birds with red breasts) and that swallows hibernate, hidden away in crevices or hollow trees. In 1703, Charles Morton, a Harvard vice president, said that, on the contrary, swallows and white storks wintered on the moon. In Persia it was thought that white storks, which have a wingspan of five to seven feet, red legs and beaks, and black feathers on the backs of their wings, annually made the pilgrimage to Mecca. An old Greco-Roman belief said that white storks didn't die of old age but flew to islands where they came to look like people. Hans Christian Andersen said these were the birds that brought babies to expectant families. It was (and still is) considered good luck if a white stork nested on your roof.

Little was known about where these distinctive, well-loved European birds spent the winter, until 1822, when one stork returned to its nest in Germany with a two-foot-long African arrow stuck through its neck. *Pfeilstorch*, it was called, "arrow stork." Over the years, around twenty-five more *Pfeilstörche* showed up in this gruesome condition.

Wikelski is an engaging public speaker, and in Copenhagen in 2018, during an informal preview of the animal internet that's been posted online, he talked about the exploits of tagged white storks. The setting was a kind of happy-hour European counterpart to a TED talk, a forum called Science and Cocktails that promises a scientific approach along with "smoky dry-ice chilled cocktails." Wikelski said, "You think we know storks, right?" It's only now, he explained, that we can learn what they do and what they know. It involves pinpointing the many places storks end up in in winter—one evaded a sandstorm in Algeria; another was eaten by a leopard in sub-Saharan Africa; and another was arrested in Egypt as a possible spy (it had been banded in Israel).

It also has to do with understanding thermals—fountains of sun-warmed air that rise a mile and a quarter into the sky and shift around a lot, depending on winds that may be pushing them sideways. Thermals don't last long, twenty minutes at most. Their true shape—transitory, drifting, and twisting—and their constantly changing speeds at different altitudes make them elusive, essentially untraceable by weather equipment. Some thermals are "virtually impossible to detect from the ground," according to an article in the *Journal of the International Space Modeling Society*.

Storks rely on "thermaling," soaring and gliding in ever-higher circles as they're buoyed up through the warm and spiraling air, which cuts down on the flapping flight that takes more effort. A flock of twenty-seven juvenile storks heading south from Germany was monitored every few seconds to see exactly where each bird was and how high up, along with their speed and direction, and how frequently each bird flapped its wings (as indicated by accelerometers). The composite pattern they formed can be seen in a color-coded video—blue swirls indicate less flapping while red means more. It's a detailed, accurate picture of a thermal, an otherwise invisible piece of the atmosphere. The birds have become meteorologists.

Some of the storks were better soarers than others. "Some were goofing off," while "some were really strong, the Schwarzenegger types," Wikelski told the Science and Cocktails crowd. The differences, he said, are so pronounced you can see who's a good flyer within ten minutes. But what researchers found really eye-opening was that this tiny span of time told you where each bird would spend that winter and the winters to come.

Follow-up studies confirmed it: storks for whom flight requires more flapping end up in Spain, more than 1,000 miles away; the great thermalers fly on and make it south of the Sahara, a 5,500-mile journey (where unfortunately, because of political conflict, they're in greater danger and may get shot). So from these first ten minutes, Wikelski said, "we can predict how far they'll go and how long they will survive." As an aside, offhandedly, he said this showed a shift in what ecology could be—a predictive science.

Groups of tagged storks that persevere, get to Africa, and evade destruction congregate in the hundreds along the edge of the Sahara. In this seemingly featureless expanse, storks have found something remarkable—exactly where desert locusts have hidden their eggs, the precise location from which the next locust outbreak will spread. If not sprayed and stopped at the source, billions of ravenous insects will eat hundreds of millions of pounds of plants in a day. How do the storks find these eggs? Unknown. Wikelski said maybe they're responding to some local sentinel, or perhaps to the scent of the eggs. Governments, he said, are taking notice. It's no longer a subject of interest only to "a crazy group of biologists."

"Martin Wikelski," says Adam Popescu in *Bloomberg Business-week*, "is the kind of guy who drives toward an earthquake." In 2016 Wikelski and his wife drove twelve hours overnight from Germany to Visso, an Italian mountain town northeast of Rome, where a magnitude 6.1 earthquake had forced the evacuation of 1,200 people. Several years earlier, he tagged goats living on the side of Mount Etna, in Sicily, Europe's most active volcano, which then had seven lava eruptions. Each time, four to six hours beforehand, the goats abruptly changed whatever they were doing. If it was after dark, they woke up and wandered around; during the day, they looked for a place of safety. Somehow they seemed to know something was going to happen. Wikelski considered the evidence strong enough to apply for a patent for a volcano-forecast system.

But what about earthquakes—could they too be known about ahead of time? Visso almost immediately experienced a second set of shocks, and Wikelski and his wife found a farmer who let them tag his cows, sheep, dogs, some turkeys, and a rabbit. "We couldn't take repeated samples actually on the turkeys and the rabbit," Wikelski said, "because they were eaten during the holidays."

The tags on the larger animals showed that yes, collectively, the Visso barnyard functioned as an earthquake warning system, with all the animals unusually active, though the amount of advance notice depended on the distance from what would be the epicenter: one hour's warning for a quake twenty miles away, and up to eighteen hours before something within six miles. Which could be time enough to save lives.

A 2019 *Science* study by ornithologists reported a "staggering loss," as *The New York Times* described it—nearly 3 billion birds disappeared from North American skies since 1970, including a quarter of the blue jays and half the Baltimore orioles. The thinness of the biosphere, it turns out, has been thinning out.

Still, in the same half century, we've been extending our awareness by tapping into the awareness of animals, creatures like tagged storks and Pluie and Deer 255. Protecting the places animals need has a new, heightened priority now that they can tell us where they go and what they face.

A Planetary Feeling

Always in the back of my mind when thinking about 50 by '50 is Benton MacKaye—someone whose thoughts roamed far and wide as he looked around, and who then reshaped the world. He had just graduated from college when, in the summer of 1900, he and a friend bushwhacked their way up Stratton Mountain,

Benton MacKaye

the highest peak in southern Vermont; the story can be found in Larry Anderson's biography, *Benton MacKaye*. There were no trails to follow, and at the summit they shinnied up tall, swaying trees to get a better view. For MacKaye this was an *aha!* moment whose echoes can still be heard.

Twenty years later, while gaining a reputation as a kind of philosopher-forester, it led him to propose building the Appalachian Trail along the ridgeline of eastern mountains from Maine to Georgia. It became the most famous trail on the planet, the hiker's Mount Everest, 2,190 miles long or, by one calculation, 5 million steps. Every year, 2 to 3 million people find their way to some part of the AT, as it's known, and it has inspired long-distance trails on every continent except Antarctica.

Now, more than a century later, MacKaye's *aha!* moment has inspired Wild East, an initiative to protect what's called the wide-open wildness surrounding the trail, an area the builders of the AT ignored but that MacKaye championed as a realm. For the Wild Easters, it's a landscape just as important as all the national parks in the West, and it can ignite a generation of Half Earth efforts.

"A damn fool scheme, Mac"

At eighty-five, MacKaye could still instantly recall that July morning of swinging from the treetops. This is from a letter he wrote in 1964 to be read aloud at a conference:

> It was a clear day, with a brisk breeze blowing. North and south sharp peaks etched the horizon. I felt as if atop the world, with a sort of "planetary feeling." I seemed to perceive peaks far southward, hidden by old Earth's curvature.

Would a footpath someday reach them from where I was then perched?

It wasn't the first time MacKaye, who grew up in a small Massachusetts town, felt the land itself changing him. Several years earlier, during an August hike to the top of Mount Tremont in the White Mountains of New Hampshire, he watched the sun rise after a night of thunderstorms—"rain coming down like 'pitchforks.'" He could see Maine in one direction, and, in another, "away in the distance," he wrote to a friend, "I could make out the hills of old Massachusetts. I felt then how much I resembled in size one of the hairs on the eye tooth of a flea." And, in his diary: "The grandest sight I ever saw was now before me, nothing but a sea of mountains and clouds."

Bill Bryson summed up his own two hundred miles of hiking on the AT in 1998 in *A Walk in the Woods,* his runaway best seller about the trail—a book that inspired so many first-time hikers it got its own nickname, the Bryson Bump. For Bryson, change didn't happen after only a single night of thunderstorms. It first came, Bryson wrote, as he looked at a four-foot-long trail map and saw that "all the effort and toil, the aches, the damp, the mountains, the horrible stodgy noodles, the blizzards, the dreary evenings . . . the endless, wearying doggedly accumulated miles— all that came to two inches. My hair had grown more than that."

Over the years the trail itself, Bryson noted, hasn't changed— unlike highways, which keep adding lanes, its dimensions remain constant, a four-foot-wide path made for walking single file: "There is the good old AT, still quietly ticking along after six decades, unassuming, splendid, faithful to its founding principles." After his hike, Bryson had changed. "I understand now, in a way I never did before, the colossal scale of the world," he wrote. "I found patience and fortitude that I didn't know I had.

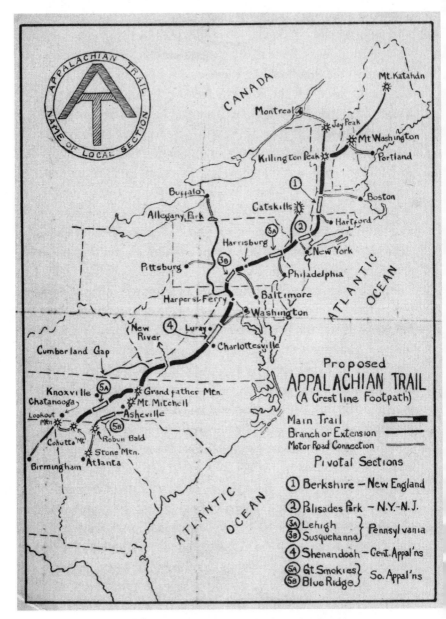

MacKaye's map for the Appalachian Trail, 1925

I discovered an America that millions of people scarcely know exists."

In 1948, Earl Shaffer, who was a radio operator in the South Pacific during World War II, became the AT's first thru-hiker (always spelled that way). Since then more than twenty thousand people have walked the trail end to end, and about four hundred have completed the Triple Crown of Hiking—anyone who's thru-hiked the AT, the Continental Divide Trail (which MacKaye helped inspire in the 1960s), and the Pacific Crest Trail, the subject of *Wild*, Cheryl Strayed's memoir about a hike that helped ease her grief over the loss of her mother.

Shaffer, carrying his army rucksack but with no tent or stove, headed north in April—"walking with spring," his own version of surfing the green wave. He reached Mount Katahdin in Maine 124 days later. He hiked, he said, to "walk the war out of my system." He later thru-hiked the other way and, at seventy-nine, repeated his original south-to-north journey. The trail, said his niece at a ceremony inducting him and MacKaye as charter members of the AT Hall of Fame, "was a flame that started in the late forties and never died."

"A damn fool scheme, Mac," said the director of a summer camp where MacKaye had been a counselor, about the idea of a wilderness trail. On the other hand, MacKaye's close friend for half a century, Lewis Mumford, the urban historian who was awarded the Presidential Medal of Freedom and an honorary British knighthood, had a profound reaction when reading about it in 1921. "I well remember the shock of astonishment and pleasure that came over me," he wrote.

I had a chance to meet with Mumford's widow, Sophia, the editor of many of Mumford's thirty-eight books, when she was ninety-three. She said MacKaye had a vivid presence of his own: "He was very crotchety. While working on an idea, he was sunk deep in it and talked about practically nothing else. He came to visit us when we lived in Queens, in Sunnyside, and said he once found his way to our house by watching the stars out the taxi window. He would read you a manuscript. And not just his own manuscript. He had a philosopher brother, and he would read *that* one, too. He was very thin and very craggy, and rough-hewn. He wore a woodsman's plaid shirts and always a red bandanna. He talked like an old Yankee farmer, full of 'ain'ts,' which was an affectation, because he was Harvard bred. He was also so lovable, affectionate, and kind, and very sweet with young people—we were younger then and considered him very old." Then Sophie, as she was called, added, "You simply felt *good* with him."

I asked her about MacKaye's influence on conservation. Sophie thought about this a moment, and said, "He was very, very close to the soil, but there was nothing of either the homesteader or the gardener in him, nothing that drew him to raise things, or scrabble in the soil. Yet in his own way he was a gardener—of a much, much larger landscape."

Optimism is oxygen

MacKaye's eye was always trained on this much, much larger landscape. In his 1921 essay introducing the idea of the trail, published in the *Journal of the American Institute of Architects*, he asks readers to see the world with new eyes, awakening a kind of Overview Effect by imagining "a giant standing high on the skyline along these mountain ridges, his head just scraping the floating clouds." Striding south through "wooded wilderness" and millions of acres of farms, the giant can see beyond the mountains to "a chain of smoky bee-hive cities" along the East Coast and to steel plants in the Midwest.

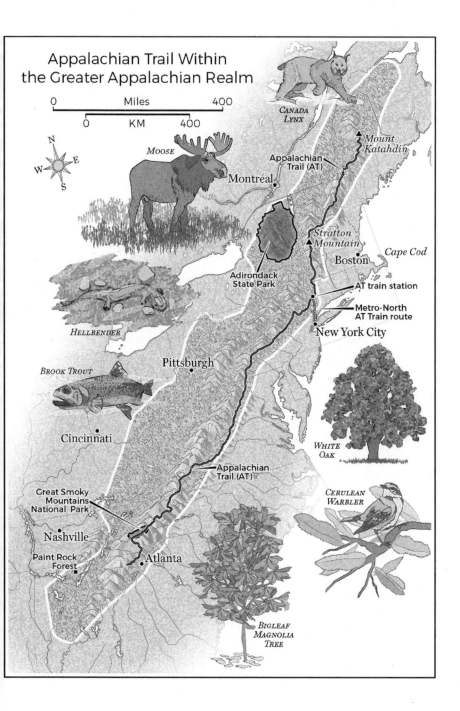

Appalachian Trail Within the Greater Appalachian Realm

Miles

0 — 400

KM

0 — 400

N W E S

Moose

Canada Lynx

Montréal

Appalachian Trail (AT)

Mount Katahdin

Adirondack State Park

Stratton Mountain

Boston

Cape Cod

Hellbender

AT train station

Metro-North AT Train route

New York City

Brook Trout

Pittsburgh

White Oak

Cincinnati

Appalachian Trail (AT)

Cerulean Warbler

Great Smoky Mountains National Park

Nashville

Paint Rock Forest

Atlanta

Bigleaf Magnolia Tree

Finally the giant takes a rest in North Carolina, on Mount Mitchell, highest point east of the Mississippi, where "he counts up on his big long fingers the opportunities" inherent in these mountains to help the people on either side by giving them "a chance to catch a breath." This is something that awaits them in the "real open—right now, this year and next." The task ahead, MacKaye says, will be "cutting channels leading to constructive achievement in the problem of living." Then MacKaye describes in practical terms the trail and the camps that could be built alongside it.

"Some mighty big things are coming out of this trail movement in the next few years if its development grows at the pace it now shows," said a piece in the *New York Evening Post* several months after the essay appeared. Over the next few years, MacKaye's "little article," as he called it, pulled hundreds and later thousands of people up into the hills to build sections of the trail in their free time, an eager outpouring of effort as unexpected as the idea for the trail itself. "A great professor once said that 'optimism is oxygen,'" MacKaye wrote in his little article.

By 1937 the entire Appalachian Trail—two thousand miles at the time—was complete, a channel through the wooded wilderness created by volunteers, though helped out here and there by young workers from the New Deal's Civilian Conservation Corps, who got paid a dollar a day. Lewis Mumford said the AT was "achieving by purely voluntary cooperation and love what the empire of the Incas had done in the Andes by compulsory organization."

It's still the case—optimism and oxygen remain undiminished. Volunteers, for MacKaye the soul of the trail, are its engineers and road crews, donating something like 250,000 ungrudging hours of work a year. The AT is in effect a national park—the longest and skinniest one anywhere. But under a unique arrangement it's also the only national park kept open by volunteers—about six thousand of them, in thirty-one Appalachian Trail Maintaining Clubs, ranging in size from the more than 100,000-member Appalachian Mountain Club to a group of students at Dartmouth College.

There are big jobs, like occasional "re-los," relocations that move the trail from one path to another, but there's also the need to keep around 165,000 rectangular white blazes freshly painted, and to clear trees brought down by winter storms. It's as if the interstate highways were built on weekends by small clubs of people who brought their own equipment. Highways sometimes have signs saying YOUR TAX DOLLARS AT WORK. On the AT, others have been here before you, preparing what you're seeing inch by inch, mile after mile. Signs on the AT might say GOODWILL— NEXT 2,000 MILES.

In 1930, Myron Avery, a Maine lawyer, took on the job of completing the Appalachian Trail, and his leadership of the volunteer groups is probably why it got finished so quickly. Quoting Harold Allen, an AT planner and volunteer, Avery said that the trail "beckons not merely north and south but upward to the body, mind and soul of man." Even so, Avery wasn't interested in MacKaye's realm; he was pragmatic and concerned only about the trail. "Our main problem," he told organizers, "is to actually create it. Then we may discuss how to use it." It's been said of Avery, nicknamed Emperor Myronides I, that he left two trails from Maine to Georgia, one of badly bruised egos, and the other the AT. For years MacKaye withdrew from all work on the trail, and his larger landscape remained neglected.

Unofficially, the Appalachian Trail has been getting longer since Avery completed it: a southern extension, originally proposed by MacKaye, the Pinhoti National Recreation Trail, terminates at a peak in the middle of Alabama. The International Appalachian

Trail (IAT) takes the trail north into Canada through New Brunswick and over the steep-sided Chic-Choc Mountains in Quebec. Then ferry rides connect to Nova Scotia, Prince Edward Island, and the west coast of Newfoundland.

The idea for the IAT occurred to Dick Anderson, a Maine fisheries biologist and later a state conservation commissioner, in 1994. "I was driving along Interstate 95, in Freeport, when it just popped into my head," he said in a 2002 *New York Times* interview. "I was so excited that I stopped at a gas station and just started talking to a guy who was pumping gas. He didn't know what I was talking about, but I was so excited I couldn't resist."

Electric with a high potential

The Appalachians—MacKaye's realm—are among the oldest mountains on earth. They were first formed 470 million years ago, and then, 299 million years ago, in a culminating event of mountain building, two supercontinents collided to form Pangaea, the most recent single world continent. This pushed up a wide mountain chain with ranges, plateaus, folds, and ridges. For millions of years its tallest peaks were as high as the Himalayas. Remnants of these Central Pangaean Mountains extend through Greenland and reappear on the far side of the Atlantic Ocean in Norway, the Scottish Highlands, and the Anti-Atlas Mountains of Morocco.

When European settlers arrived, the idea of the Appalachians as a single place began to fracture. "The circumstances surrounding the naming of Appalachia," wrote historian David Walls, "are as hazy as a mid-summer's day in the Blue Ridge." It may trace to sixteenth-century Spanish conquistadors in Florida who heard rumors that to the north there was gold in a region called Apalachen.

In the nineteenth century some writers preferred to call the mountains, or at least the northern half of them, the Alleghenies, a French spelling left over from pre–Louisiana Purchase days and still the name of a river in Pennsylvania. Washington Irving thought that to create a national identity the whole country should be called either the United States of Appalachia or the United States of Alleghania. "I should prefer the latter," he added. There are still regional pronunciation differences. People in the North say *Appa-lay-chin;* people in the South say *Appa-latch-in.*

In 1873, Will Wallace Harney, an Indianan who moved to Florida and spent much of his life writing booster pieces persuading northerners to move to Orlando, wrote a short story about Appalachia, "A Strange Land and a Peculiar People." As David Walls notes, this was the first installment of decades of "local color" magazine stories and essays written for readers on the East Coast and in the Midwest by a succession of writers, a college president among them, about the Appalachians and the people there, particularly those in southern Appalachia, which inescapably comes across as a place of "otherness."

Go there, one writer declared in *Harper's Magazine* in 1886, and "you detach yourself from all that you have experienced." The college president William Goodell Frost of Berea College said in *The Atlantic Monthly* in 1899 that mountain people had just awakened from a century-long Rip Van Winkle sleep and should be thought of as our contemporary ancestors.

More chillingly, Harney's story introduced the themes of degeneracy and degradation, suggesting—with no basis in fact— that Appalachian people showed "marked peculiarities of the anatomical frame" such as "elongation of the bones" and "loose muscular attachment of the ligatures." (Equally baseless accusations have been made about Indigenous communities, African Americans, and others.)

A hundred years later, in the 1970s, sinister hillbillies were central figures in James Dickey's book *Deliverance;* the film still haunts local communities. More recently, in 2016, there's *Hillbilly Elegy* by J. D. Vance, a best-selling memoir that reiterates the idea that many problems in Appalachia are the fault of its residents. Throughout this long and continuing trend, the outside world, primarily city dwellers, had no problem denigrating the region while extracting its coal and leveling its forests, and more recently fracking it and draping it with oil pipelines.

For MacKaye, the champion of the Appalachian realm, it was a place of strength and sanity, and the people most at risk of becoming stunted and benighted were the city dwellers on either side of it, the same people who now might never get to see it. Actually *everyone* would soon be at risk because, MacKaye wrote in his impassioned book *The New Exploration,* a crisis was approaching. The realm itself might disappear or be disfigured beyond recognition.

With the coming of highways and electric power lines, cities and suburbs were explosively spreading outward like an incoming army, imposing a new kind of wilderness on the landscape—a "jungle of industrialism," a "creeping labyrinth," an "iron cobwebbing," a "standardized excrescence." Appalachia was seeing the approach of forces "far more terrible than any storms encountered within uncharted seas." Yet the realm might still be rescued. Appalachia, MacKaye said, "is electric with a high potential—for human happiness or human misery."

The New Exploration, which reads like a charter for the landscape, was published in 1928, seven years after the Appalachian Trail essay. It has been in and out of print ever since, passed from hand to hand when not otherwise available—discovered, forgotten, rediscovered. Lewis Mumford considered it a classic, writing a preface in the 1960s: "*The New Exploration* is a book that deserves a place on the same shelf that holds Henry Thoreau's *Walden* and George Perkins Marsh's *Man and Nature;* and like [*Walden*], it has had to wait a whole generation to acquire the readers that would appreciate it."

Mumford makes the point that MacKaye is speaking older truths to address the problems of his time, and, it turns out, of ours as well. "In Benton MacKaye," Mumford went on,

> the voice of an older America, a voice with echoes not only of Thoreau, but of Davy Crockett, Audubon, and Mark Twain, addresses itself to the problem of how to use the natural and cultural resources we have at hand today without defacing the landscape, polluting the atmosphere, disrupting the complex associations of animal and plant species upon which all higher life depends, and thus in the end destroying the possibilities for further human development. That voice was needed in 1928; and because it was not listened to, it is needed even more today.

To be a new explorer, then, is what's required to move forward with a Half Earth point of view. MacKaye's vision is a charge to think big, see more widely, take it all in. "The new explorer of this 'volcanic' country of America," wrote MacKaye, "must first of all be fit for all-round action: he must combine the engineer, the artist and the military general. It is not for him to 'make the country,' but it is for him to know the country and the trenchant flows that are taking place upon it." MacKaye's motto: "Speak softly but carry a big map!"

Like many planners of his time, MacKaye saw humans as the beneficiaries: "The primary object of the Appalachian project is human," he wrote in a 1922 essay. "The human biped comes first." At the same time *The New Exploration* offers a fresh calculation of what gets lost in people when landscapes disappear, and what is gained when landscapes are kept intact. While earlier writers emphasized the importance of a pure wilderness above all ("The clearest way into the Universe is through a forest wilderness," said John Muir), MacKaye saw three types of landscapes, which he called "elemental environments," all interconnected—the *primeval* (wild landscapes, "the environment of life's sources"); the *rural* (farms and villages); and the *urban* (cities).

Once people had ongoing access to all three elemental environments, a new awareness kicked in, something enduring and indelible. In his math: $1 + 1 + 1 = 4$. This was, MacKaye wrote in *The New Exploration*, "not so much an affecting of the countryside as of *ourselves* who are to live in it."

He called it "a higher estate in human development" and "the gradually awakening common mind" that would let people live as "a unit of humanity" with the next generations. This was the planetary feeling he'd had on the summit of Stratton Mountain in the summer of 1900 at the age of twenty-one, the feeling that pulled him ever forward from that day on.

Voyages of discovery about home

"These Waverly Oaks are, all things considered, the most interesting trees in eastern Massachusetts," wrote Charles Sprague Sargent in 1890; he was the director of Harvard's Arnold Arboretum for more than fifty years, where he assembled between five and six thousand species and varieties of trees and shrubs from around the world. The little stand of giant white oaks he was referring to, in Belmont, near Boston, was the remnant of an ancient forest. Sargent, the foremost tree expert of his day, said some people thought the smallest of the twenty-three oaks was more than one thousand years old, but he considered this greatly exaggerated. He thought it far more likely they were between four hundred and five hundred years old: "It is safe to surmise that the youngest of them had attained to some size before the Pilgrims landed."

Moreover, Sargent said, they were in good shape—"still healthy, and are growing with considerable vigor" and "may continue, with proper care, to live and increase slowly for centuries." Winslow Homer had painted their extraordinary shapes, immense but gnarled, towering over the woods around them. Longfellow had his portrait painted beneath them. But now there was a railroad station within a few hundred yards, and "the whole region is undergoing rapid development, and houses are springing up on every side," Sargent wrote. He suggested that a small public park of no more than three or four acres "would protect the trees from the dangers which now threaten them."

Waverly Oak, 1924

It might have ended there, but a subsequent issue of *Garden and Forest*, Sargent's magazine, carried a follow-up letter from a young landscape architect, Charles Eliot, son of Charles W. Eliot, Harvard's president for forty years who had a mythic reputation at the college. Karl Haglund, a historian and urban planner, noted that it took some years for the younger Eliot, plagued by self-doubt and depression, to find himself. Growing up, Eliot went on long hikes around Boston and drew maps of what he saw (MacKaye also made maps as a child). But then Eliot met Frederick Law Olmsted, became his apprentice, and two years later set up his own landscape-architecture practice, designing parks in New Hampshire, Ohio, and Salt Lake City before teaming up with Olmsted's sons. Eliot believed that special natural landscapes were the cathedrals of the modern world.

Eliot's letter, "The Waverly Oaks: A Plan for Their Preservation for the People," was the beginning of seven years that transformed the Boston area into what Haglund calls the "Emerald Metropolis." This was an expansion of the Emerald Necklace, a linking of city parks Olmsted had worked on. Eliot's project saw the creation of more than nine thousand acres of parkland (including the Waverly Oaks) within ten miles of the statehouse in downtown Boston. It was landscape planning that took the idea of setting aside natural wonders, like Yellowstone or Yosemite, and applied it to metropolitan areas.

The letter was a kind of master map of what remained of the primitive wilderness of New England, near Boston. "Reservations," Eliot called these places, and he found them along the coast, on beaches, riverbanks, and wooded hilltops, and connected them all by parkways. Exploring the Emerald Metropolis, a colleague of his later said, was like going on "voyages of discovery about home."

Having survived so long, many of these places were now "in

daily danger of utter destruction," Eliot went on. It would take a regional parks commission to protect the larger spaces, but he proposed "an incorporated association, composed of citizens of all the Boston towns, and empowered by the State to hold small and well-distributed parcels of land free of taxes, just as the Public Library holds books and the Art Museum pictures—for the use and enjoyment of the public."

Eliot's idea, which he called "an imperfect outline of a scheme," took form as the Trustees of Reservations, the world's first land trust (a nonprofit group that protects natural land and farmland). Since then, working only in Massachusetts, it has protected more than 27,000 acres. In 1893 the state set up the Metropolitan Park Commission—and right away it created a fifty-nine-acre reservation that included the Waverly Oaks. Which meant there were now both landscape libraries and landscape cathedrals in the Boston area—which might, Eliot said, also be thought of as the fountains that could slake people's thirst for the countryside. Eliot died young, at thirty-seven, from spinal meningitis. Today, in the United States alone, land trusts have protected 56 million acres, and counting.

Only one of the stunning Waverly Oaks still stands. As for the rest of them, they got loved to death—so many people came by train and trolley to admire them, the soil compacted, the roots got trampled. I had a chance to see the sole survivor with conservationist James Levitt, who lives in the area and works in Cambridge at the Lincoln Institute of Land Policy.

First, you pass by descendant oak trees, already sizable but not yet majestic, on a nearby hill—Sled Hill to neighborhood kids. Local citizens, led by Levitt and a couple of land trusts, have created an elegant pathway to the great tree, the three-quarter-mile-long Waverley Trail (now spelled that way), most of it on the sidewalk of a busy street that passes Waverley Square, better known as Car Wash Corner. The trail has a green stripe down the middle, banners, bronze medallions, and kiosks celebrating the history of the town and the trees.

The oak, as you approach it, seems tucked away—behind a lawn, set back from a parking lot, and down near a stream that in summer has croaking frogs. This out-of-the-way spot might be why it's still alive—and if it was five hundred years old in Eliot's day, it's now at least six hundred.

Surprisingly, it's as quiet as a library under the tree, and in a breeze the branches overhead shift and the leaves shimmer, like water bubbling in a fountain. Standing with my back to the great white oak, I looked out beyond the parking lot at Belmont, a 160-year-old suburban town. In that moment, it seemed like a town in its infancy, since it was honoring a commitment to protect something that may need a thousand years to complete its lifespan. So maybe saving one old tree isn't setting aside half the earth for the rest of life, but simply standing beneath it can summon a planetary feeling.

Tenuous and vulnerable

There's a little trail in the making, a mini-AT, also called an outer Emerald Necklace, halfway between Eliot's reservations and MacKaye's countryside—a stepping stone to 50 by '50 because of the connection between people's deep feeling for the land and the desire to protect it. It's the Bay Circuit, 230 miles long, a tenth the length of the AT, and it runs in a semicircle around Boston, starting and finishing on the shores of Massachusetts Bay.

Like the AT, it has white markers and is kept up by volunteers. (One trail marking tip: "When nailing markers to trees, leave 3/4″+ of exposed nail to allow for tree growth.") It's not a ridgetop

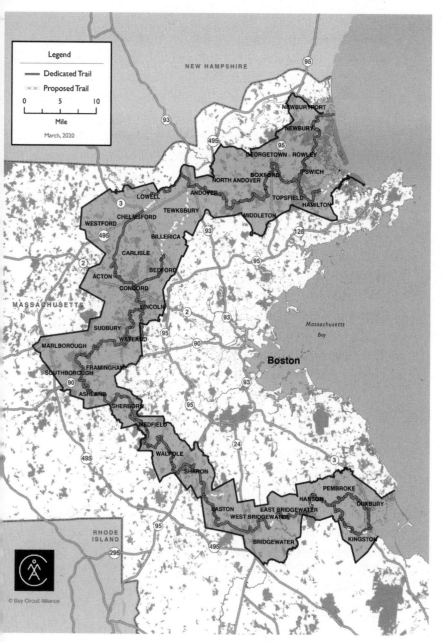

The Bay Circuit

pathway—there's no chain of mountains surrounding the city it could follow. In the north, it climbs what hills there are; some have long views to the Boston skyline twenty-five miles away. Historically, much of the flatter land in the south was used for cranberry bogs. The biggest one was harvested by Ocean Spray. That particular bog is now a protected area, a haven for birds and for turtles such as the northern red-bellied cooter, and it's reverting to what it had been, an Atlantic white cedar swamp. The trail runs across the swamp on "the Tunk," as the local Wampanoag tribe once called it—the "Indian Crossway" to colonial settlers— a mile-long, straight-line, elevated causeway built by Native Americans and used by them for a thousand years.

MacKaye had a hand in scouting locations for the Bay Circuit in the 1920s and 1930s, but it wasn't his idea. You could say it belonged to Alice Buck as much as to anybody, because of what she did to save Indian Ridge.

This steep glacial hill had been by far the most important piece of land in Andover, north of Boston. Doctors advised patients to walk its paths daily. In the 1890s, Squire Samuel Farrar, the town's leading citizen, who lived to be ninety-one, credited the hill for his long life. "I saw all my own wood," he said, according to *The Andovers* magazine. "I work in my garden an hour every day, and I have walked twice a day around Indian Ridge for 50 years." An Andover historian, Juliet Haines Mofford, quotes a teenage mill worker from the 1820s: "It was everything for us to have Indian Ridge close at hand. We were always going there for the first flowers, or May morning walks, or to get a drink at the Red Spring, or clear through over the ridges berrying. We had picnics, temperance lectures. . . . Whatever was wanted out of doors, we had our place for it."

In 1896, the owners of twenty-three acres on the ridge planned to sell their land so it could be logged and then leveled for the sand and gravel belowground. They wanted $4,000, the equivalent of more than $120,000 today. To raise money, Alice Buck, an amateur botanist, and a few friends from the town, wrote articles that got published in New York and Boston newspapers, and contributions poured in, one from Germany. The Andover Cricket Club raised $4.10. Elaine Clements from the Andover Center for History and Culture says the community was able to come up with $3,500, which was accepted.

The successful campaign was commemorated in 1907 by a large bronze plaque, twice cleaned and restored since then and still firmly bolted to a massive boulder on Indian Ridge that you can walk past on the Bay Circuit: "In memory of Miss Alice Buck, by whose loving interest and untiring exertion the perpetual use of this woodland by the people of Andover was secured in 1897."

The name "Bay Circuit" got coined in the 1920s—various people claimed credit for it. The idea was that the state would start buying reservations of low economic value but with great scenic or recreational possibilities. The way forward seemed clear. Larry Anderson, Benton MacKaye's biographer, references a 1950 article by Henry M. Channing, a Boston lawyer and member of the Governor's Committee on Needs and Uses of Open Spaces. Channing, who recalled looking at a map of existing public spaces, was "amazed to see that many of the outstanding features—of wide variety—fell within a crescent-shaped arc." Tied together "through great stretches of attractive country" they could become an eastern Massachusetts circuit of parks.

But this ambitious 1929 plan reached the governor's desk a few months before Black Tuesday, the day of the stock market crash, and the state didn't fund the program until the 1980s. By then even more of the potential reservations identified in the 1920s had been protected by a dozen towns, the way Indian Ridge had been in the 1890s. Which meant that, however negligible the economic value of these hills and woods and marshes and river valleys, they were deeply important to people. But there was no linkage from one town to the next: emeralds but no necklace.

In 1930 MacKaye proposed adding "an earthy footpath" along the length of the Bay Circuit, but no one took him up on the idea. When the state ran out of money again at the end of the 1980s, the Bay Circuit might have been permanently short-circuited, except that in 1990 Alan French got involved—"in my small way," he said to me during a hike to Indian Ridge. Over the next twenty-two years, before stepping aside at age eighty, French got MacKaye's long-forgotten trail built, mostly by hundreds of volunteers.

People have struggled to describe how French pulled this off as an unpaid worker for a tiny nonprofit called the Bay Circuit Alliance. It was like the AT all over again, only better, it was said, because French combined MacKaye's sweetness with the drive and persistence of Myron Avery. Going from town to town and person to person, French would find people eager to do the work and would then make a point of giving them all the credit. "There are Alice Bucks in every town," French told me. "You just need to find them and empower them. Every year we built ten or fifteen more miles, and by 2000, the trail had legs. So to speak."

At one time, French was a publishing executive, business manager of *McCall's* in New York, a magazine with 8 million readers. In midlife French decided what he really wanted to do was stay close to the outdoors. "To have a life that is small, with small organizations," he said on our hike. He opened a camping and backpacking store in a small town, which happened to be Andover, where a visiting reporter found the headquarters of the Bay Circuit Alliance and described it as "a cluttered corner of piled papers and file folders that looks like it was organized by pack rats."

The Bay Circuit's not yet quite complete; maps have dotted

lines for the small gaps. It's within driving distance of 4 million people, but so far only about a hundred have become thru-hikers—French among them (he's walked it six times). Like the AT in its early days, it's taking time for the Bay Circuit to get noticed. Unlike the AT, it cuts through towns that are becoming suburban. Like the AT, it works to open minds to the shapes of a region, in this case connections to Boston Harbor and out to the Appalachian Trail itself, in the northwest corner of the state.

There are several semicircles around Boston now—two of them are beltways about twenty miles apart (Route 128 and I-495) with the Bay Circuit in between. Some people see the situation as a test of whether, over time, the trail will have the gravity to rally people to anchor the countryside that remains. In 2019 one town saved twenty-four acres of forest close to the Bay Circuit that a developer wanted. It was like the Indian Ridge fight all over again, victory included. French, in his eighties, has founded a new group, even smaller, to protect even more of the countryside, and he sees dozens of special places that could become new reservations—adding pearls to the emeralds, he says. (He adds, "Don't let the trail wag the dog.")

How much of a planetary feeling can you get along the Bay Circuit? One thru-hiker, Dan Brielmann, a community activist and videographer, told me he remembers only a series of disappointments—a hilltop had beckoned, but when he got there, the fire tower that should have been his lookout was being used as a cell-phone tower and was covered in barbed wire.

Larry Anderson, also a thru-hiker, told me, "A planetary feeling? Sure, but it's a balancing act. In the towns, you're walking through other people's activities. Beyond the towns, there were days I had the land pretty much to myself, more alone than I would be on the AT in the summer. I got lost a few times—which in itself was something, that there were areas big enough to get lost in. But the experience of going in and out of the wild—it was like getting a strong signal followed by a weak signal. It made the trail feel tenuous and vulnerable."

Ownership

William Wordsworth, who became poet laureate of England, spent sixty years in that country's Lake District, more than half a million acres of long, deep lakes, heather-blanketed mountains below bare, craggy peaks, and valley-bottom farms, pastures, and villages; it's now a World Heritage site. Midway through his life, in 1810, Wordsworth needed money and wrote *A Guide Through the District of the Lakes in the North of England*. Then, as now, most land there was privately owned. All the same, Wordsworth said, he and many others could "testify that they deem the district a sort of national property, in which every man has a right and interest who has an eye to perceive and a heart to enjoy."

Right after World War II, the new Labor Party government pledged to win the peace by making Britain a country worthy of its fighters, a "land fit for heroes." Wordsworth's idea of the countryside as a sort of national property was officially adopted: "We regard the countryside as the heritage of the whole nation." G. M. Trevelyan, then one of the foremost English historians, said, "Without natural beauty, the English people will perish in the spiritual sense." The National Parks and Access to the Countryside Act, adopted in 1949, was called a "people's charter for the open air."

There are now fifteen national parks in England, Wales, and Scotland, including the Lake District, welcoming around 75 million visitors a year. They're a mix of wildlands, farms, and villages—the primeval and the rural, in MacKaye's language—

and it's part of English law that they are "the most beautiful, spectacular, and dramatic expanses of country." Unlike U.S. national parks, they're "greenline" parks. This means they're permanently protected but owned, for the most part, by the people who live there—on farms, in villages, and on big estates—the custodians of the countryside.

Then there are the forty-six Areas of Outstanding Natural Beauty (AONBs, a name acknowledged to be clumsy). These too are greenline parks, mostly farmland, and found in lowland areas with little wild land left; they're described as gentle rather than dramatic landscapes. But because of greenlining, more than a third of England is either a park, or an AONB, or part of the "green belts"—rings of open land surrounding fourteen cities off-limits to development to prevent sprawl. This means a tremendous amount of land will remain unbuilt, assuring the future of "beauty that took three hundred years to grow, and can never be replaced," as the novelist E. M. Forster wrote in a 1938 essay.

Greenlining

Long ago I came across a faded, large, and rather sketchy hand-drawn map of the United States, bigger than anything MacKaye or Eliot ever drew, far less elegant. The lines on it, originally purple, the ubiquitous color of copies before xeroxing, are now a dark brown. It has a first-draft, good-enough-to-be-pinned-up-while-we-refine-it look to it, with no orienting information, such as cities, mountains, rivers, highways, parks, or forests. All it shows, aside from state lines, is roughly two hundred irregularly shaped areas spread out across the country, each indicated by dotted lines—some roundish, some long and thin, others lumpier and more paramecium-like. Almost all are unfamiliar despite neatly

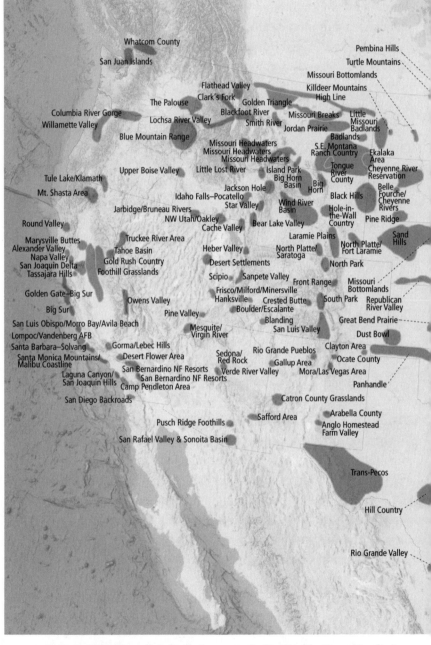

Potential U.S. Greenline Parks (map newly digitized by Dana Tomlin)

Red River Valley

Aroostook Valley

Peatlands

North Shore

Keweenaw Peninsula

Maine Coast

Iron Range

Apostle Islands Shore

Lake Champlain & Valley

White River Valley

Penobscot Bay

North Shore/Apostle Islands

Adirondacks

Maine Coast

Sheyenne River Valley

Northern Highlands

Thousand Islands

Limestone Valley

Lake Winnipesaukee

Connecticut River Valley

Prairie Pothole

Door Peninsula

Grand Traverse Area

Tug Hill Plateau

Merrimack River

Minnesota River Valley

Saginaw Valley

Mohawk River

Berkshires

Cape Cod

Mississippi Blufflands

Finger Lakes

Catskills

Nantucket Islands

Driftless Area

Kettle Moraine

Fruit Country

Central Highlands

Hudson River

Housatonic

Narragansett Bay

Missouri Bottomlands

Susquehanna

Poconos

Delaware

Eastern Long Island

Amish Settlements

Amish Country

River

Pinelands

Loess Hills

Amish Settlements

Brandywine

Amish Settlements

Tygart River Valley

Great Valley

Delaware Bay Estuary

Van Buren/Lee Counties

Hocking Hills

Canaan Valley

Eastern Shore

Illinois River Valley

Piedmont

Flint Hills

Amish Settlements - Ohio River

Daviess County

Counties

Adams County

Mennonite Area

Crawford Uplands

Knobs Region

Greenbrier River Valley

Ohio River Counties

Cumberland Plateau

Little Balkans

St. Francis Mountains

Shawnee Hills

Bluegrass Region

Clinch/Powell Watersheds

Irish Wilderness

Reelfoot Lake

Pamlico-Albemarle Peninsula

Buffalo River

Salem Plateau

Cherokee/ Adair

Crowley's Ridge

Hatchie River Floodplain

Sequatchie Valley

Jackson County

Lookout Mountain

Ouachita Mountains

White River

Chattahoochee River

Cross Timbers

Warm Springs/ Callaway Gardens/ Pine Mountain

Low Country

Ogeechee River

Red River

Mobile-Tensaw Bottomlands

Altamaha River

Pearl River

Pineywoods

Pascagoula River

Suwannee River

Atchafalaya Basin

Apalachicola Bay/Estuary

Kissimmee River Fish-eating Creek

South Florida/Everglades

Florida Keys

typed names. Several are enormous while others are more modest in size; some bump up against one another, though many are
isolated; long squiggles might be river valleys. Each state has two
or three, and altogether they cover about a quarter of the country.

This map was one of those things that caught hold of me and
wouldn't let go. Over the years I showed it around, and I was told
this was something a group at the U.S. National Park Service had
worked on. But why was the Park Service singling out this vast
extent of little-known land? Nothing more than a hodgepodge,
it seemed, a mass of blobs, no pattern to it. I must've written to
someone in the government about it, because in 1994 I got a letter from J. Glenn Eugster, then a watershed program manager at
the Environmental Protection Agency. Eugster dated the map, or
at least the idea behind it, to the 1970s.

From Eugster's letter: "The map has been called different
things at different times. Originally, it was a list of 'potential
Greenline Parks' [then] 'Great Landscapes of America,' and
[finally] 'Potential Protected Landscapes.'" The blobs, it turns
out, are the landscapes people adore. In other words, a treasure
map. Which makes it an Americanized version of the English
national parks and AONBs—though it echoes something even
earlier, and American in origin.

Back in 1890, greenline parks in the United States were
known, confusingly, as blue-line parks, because the first map
of the very first one, Adirondack Park in upstate New York, had
a line around it drawn in blue ink. Still the shining example of
American greenlining, this 6-million-acre park with one hundred
peaks and three thousand lakes is the largest state park in the
Lower 48, three times the size of Yellowstone. Twice it was the
site of the Winter Olympics, and 7 to 10 million people visit it
every year. Yet more than half is privately owned. Under Article XIV of the New York State Constitution, anything the state

buys within the Adirondack blue line (it's still called that) must "be forever kept as wild forest lands." A state commission set up in 1971, the Adirondack Park Agency, steers any growth toward existing towns.

The hand-drawn map of many names was a way of outlining a second wave of national parks, American AONBs, special places the government could never buy all of, or that were too small, or had too much farmland, or weren't wild enough to become national parks. In 1978, Congress authorized the first of these "national reserves," the name it finally settled on, in the New Jersey Pinelands.

Here was nearly a million acres in the most urban state still filled with forests, cranberry bogs, and blueberry fields, and sitting on top of an aquifer with water quality that, it's said, is nearly as pure as melting glacier ice. Half the land is privately owned. Local people help make decisions about any growth, something that's true in English parks as well. A decade earlier, the Pinelands was proposed, and rejected, as the site for the biggest supersonic jetport in the world, plus a new city for a quarter million people.

After the Pinelands, the next step was supposed to be a countrywide system of national reserves. It would by definition be a MacKaye-sized vision, since it could protect far more land than the existing national parks or wilderness areas. It was to be the signature environmental project and lasting legacy of the second Jimmy Carter administration. Only there never was one. The first Reagan administration took its place and turned away from the plan. Since then, only one other national reserve has been created, in 1988—the City of Rocks, 14,400 acres in Idaho with five-hundred-plus-foot-tall rock skyscrapers and rock spires, steeples, arches, and tafoni (caverns that weathering has carved out within steep cliffs). The City of Rocks is also called the Silent City of Rocks.

A huge-enough place

Ann Satterthwaite, a Washington, D.C.–based environmental planner, has been thinking about the future of Benton MacKaye's Appalachian realm at least since 1974, when she wrote "An Appalachian Greenway: Purposes, Prospects and Programs." This was a proposal to fulfill MacKaye's "wild dream," as she put it, of protecting 28 million acres of countryside. MacKaye was still alive, although at ninety-four too frail for a visit. Satterthwaite remembers he was very excited and enthusiastic when they spoke on the phone. "It comes to me as a delightful surprise," he wrote to her about her report, saying that from the very first moment he began thinking about the trail, he considered a greenway "as needful to the protection of the primeval environment" as the trail itself. "I have many times tried to make this point but not too successfully. But your brochure makes the point in no uncertain terms."

So they never met—but to people in the Wild East initiative, dedicated to protecting sweeping vistas and landscapes beyond the trail, Satterthwaite is the continuity, the direct line to MacKaye's original vision, and her report (which MacKaye called a brochure) is no longer an obscure, mimeographed document.

In the 1970s, the AT was going through problems familiar to anyone walking the Bay Circuit today. Officially, the AT was a congressionally designated national scenic trail, but 40 percent of it ran through private land, and 10 percent was still routed along roads, streets, and sidewalks. Even in the woods people were putting up summer homes practically next door, turning parts of the trail into, as Satterthwaite described it, "a narrow path winding through second-home developments, with background noises of chain saws and barking dogs."

Satterthwaite, who worked in both the United States and England and is a former chair of the National Association for

Ann Satterthwaite

Olmsted Parks, had a novel suggestion. In the center of the greenway, there would be a wilderness-type zone. It would be publicly owned and kept in a "primeval state." Beyond, on either side, would be countryside zones extending up to ten miles. These would be privately owned, and in these zones, she wrote, "life and activities will proceed as usual." As long as the greenway is respected, she made a point of insisting.

Essentially, the AT would be a national park wrapped up in an AONB. Satterthwaite's AONB idea went nowhere at the time, but her report stirred the federal government to start buying off-road land for the trail, along with a permanent wild zone about a thousand feet wide.

Looking back, Satterthwaite's only regret is not pushing for an even bigger countryside zone, maybe thirty or forty miles wide. "I shouldn't have given up so easily," Satterthwaite went on. "There was a lot of support among people I spoke to, but the highest priority for the hikers was protecting the trail itself." This time around she's more direct, and the Wild Easters, who gather annually as the Appalachian Trail Landscape Partnership, are used to hearing her tell them to do more—even when they talk about thinking a century ahead, or protecting a million acres.

Satterthwaite eloquently told me, "The Appalachians, historically a great divide, are the potential Central Park of the East, two thousand miles long and one hundred miles wide, which puts them on a par with the big landscapes of the West. Like Central Park, they have a lot of people on either side—in 'beehive cities,' as MacKaye called them. The Appalachians are the backyard of a line of eastern cities, but also the front yard of a line of midwestern cities, like Pittsburgh and Cincinnati. I didn't get into environmental issues back then, the fact that because of global warming plants and animals will be making their way north through the Appalachians. But in MacKayean terms there's also an abiding planetary need for the mountains to reach deep into the lives of city people. Already we have to think bolder than the Wild East—bigger is always better—and include the Tamed East, the cities flanking it. Now we're talking about pretty much all the land between the Atlantic Ocean and the Ohio River, a huge-enough place. From sea to shining river."

Twice a day from Manhattan

So far, Manhattan is the one place where the Appalachians reach deep into city life, getting as far as Grand Central Terminal in the heart of Midtown. It's the only direct mass-transit link with the AT, the kind of connection Ann Satterthwaite calls a "green finger." Every weekend and on holidays, two morning trains leave Grand Central (the crossroads of a million lives, an old radio show called it), and just about two hours and sixty-six miles later the trains stop at a short wooden platform made of two-by-six planks that's just long enough for a single door on a Metro-North commuter train to open up and let hikers off. Later in the day, three trains offer a return trip. (Service was suspended during the COVID-19 pandemic.)

The Appalachian Trail Station, a couple miles north of the town of Pawling, New York, also has a bench and a bulletin board, both painted hunter green, and there's a low railing, no roof, and a trash can. For a time, there was a birdhouse on top of the bulletin board. The railroad installed a large metal sign that says "Appalachian Trail," and over the bench there's a smaller unobtrusive brown wooden trail sign that tells you Connecticut is several miles away. The station, *The New York Times* reported, cost ten thousand dollars to build, and once hosted a wedding, and exists because a former president of the railroad is an avid hiker. As soon as you get off the train, crossing the single line of track just south of the station, the trail is right there.

Even after the AT wild zone was secured, there used to be worries that the section of the trail near New York City would seem hopelessly suburbanized, serving only as a bridge or connector on the way to wilder landscapes north and south. But an eighteenth-century sense of wildness still clings to the area, and it feels immensely far away from twenty-first-century city life.

This is a legacy of the Oblong, territory argued over by colonial New York and colonial Connecticut—a strip of land two miles wide and sixty miles long that stayed unsettled and unowned until Quakers arrived in the 1730s. In 1767, almost one hundred years before the Emancipation Proclamation, local Quakers abolished slavery among their members, reasoning, it was said, that God is present in every person and you can't enslave God.

Paralleling the Oblong is the Great Swamp, another strong-

hold of wildness, this one 25,000 years old. It's a twenty-mile-long wetland, and its six thousand (mostly hidden) acres are a source of drinking water for a million people and a home for otter, beaver, mink, bobcats, and around 180 species of birds. A rare, recently discovered leopard frog species is found in the swamp, and it's not unusual for some five thousand ducks to show up at a pond in the middle of it on a single evening.

The Great Swamp has its own dedicated group of supporters—FrOGS, Friends of the Great Swamp. Seventy-five AT volunteers have created the best way to see it, a long, gently meandering boardwalk that crosses the swamp and seems to float above it. Past the swamp the trail heads up a forested hill. Down the other side is the Dover Oak, 114 feet tall, the largest oak on the AT. A white oak like the Waverly Oaks, the Dover Oak has a twisted trunk that's even more massive, and, though at three hundred years old it's younger, it represents a living link to the old Oblong wildness.

For me, the AT train stop is about as good as a trip to the International Space Station, although clearly that's not the kind of comment you'd expect to find on a ridership satisfaction survey.

But I'd call it a Half Earth induction portal or biosphere reentry point. A century ago Benton MacKaye laid claim on our behalf to the vastness of the eastern landscape, and since then four generations of hikers have built and maintained the trail from peak to peak that he could already see in his mind.

In keeping this wildness intact and nearby, the trailmakers are also ushering us into a presence older than mountaintops or continents. It's our marvelous privilege to be participants in and guardians of an ancient community billions of years old, the continuousness of life itself.

Beyond the Dover Oak, the AT climbs another hill to Cat Rocks, a vista point with a panoramic view of the Oblong. It's a preview of what Wild East wants to accomplish. Within this view is the Boniello tract, 219 acres of woods where for years a developer wanted to build fifty homes—until a fund-raising campaign made it part of the Appalachian countryside. The grand view from Cat Rocks showcases this permanently protected piece of MacKaye's realm.

Staying started

Ole Amundsen, an environmental consultant based in Maine, always puts these questions to people who are thinking about protecting a place: "Who's the Thoreau of your area? Who's the Andrew Wyeth?"

When John McPhee, then a young *New Yorker* reporter, wrote *The Pine Barrens,* a 1968 book about his affection for the New Jersey Pinelands, he assumed he was writing an obituary. He wasn't a Piney, as people from the Pinelands are called, but he'd grown up not far away in Princeton. "I wanted to see it while it was still there," he said.

McPhee began exploring what he called "this distinct and separate world," which seemed to have fewer people in it than before the American Revolution. Often, shifts were abrupt: "It is possible to be at one moment in farmland, or even in a residential development or an industrial zone," he wrote, "and in the next moment to be in the silence of a bewildering green country." Once inside, there were hills you could climb, and no matter which way you looked, "in a moment's sweeping glance, a person can see hundreds of square miles of wilderness."

McPhee found the Pinelands as incongruous as they are beautiful. It was so unlikely, he thought, to find so much unbroken forest so near big cities like New York and Philadelphia. McPhee suggested making the area a national reserve but closed his book with the assumption that the place was doomed:

> Given the futilities of that debate, given the sort of attention that is ordinarily paid to plans put forward by conservationists, and given the great numbers and the crossed purposes of all the big and little powers that would have to work together to accomplish *anything* on a major scale in the pines, it would appear that the Pine Barrens are not very likely to be the subject of dramatic decrees or acts of legislation. They seem to be headed slowly toward extinction.

But sometimes a book can enter the debate—McPhee's became a best seller—and reach a reader who can obliterate futilities. Brendan Byrne, governor of New Jersey from 1974 to 1982, played tennis with a group that included McPhee. Byrne read McPhee's book and told McPhee he was determined to prove him wrong. Byrne wasn't a Piney either; he'd grown up outside Newark. But his interest in the Pinelands was directly responsible for the area becoming a national reserve, and it's the accomplishment Byrne was most proud of. He gave the credit to McPhee: "If there's one person without whom there wouldn't be a Pinelands Act it would have to be John McPhee."

For Carleton Montgomery, executive director of the Pinelands Preservation Alliance, a watchdog group, it isn't getting started that's so crucial, it's staying started, in a way. "Every battle you lose is forever, and every battle you win is provisional," he told me. On the other hand, he sees the Pinelands as having a resilience that is very hard to overcome. A five-year struggle led to a twenty-two-mile natural-gas pipeline *not* getting built through protected land. The greenline boundary around the Pinelands only protects about two-thirds of its ecosystem, but even so you can do what McPhee did more than fifty years ago—"see it while it was still there." Montgomery's group organizes an outing he calls the "John McPhee Pine Barrens Today Tour." There are still dirt roads of white sand that glow in the moonlight, and in the streams there's still cedar water—reddish water the color of bourbon because of iron in the soil and tannins that leach from the cedar trees.

The boundlessness has survived, too. One day Montgomery drove me through only a fraction of the reserve's more than eight hundred thousand acres of woods. For several hours we passed no other cars, mile after mile. Looking as far we could see in all directions (à la McPhee), there were no hints the forest would ever end. We went down a sandy road to a spot Montgomery particularly likes and came to a bridge over a stream; on the far side was a long-abandoned village.

There were no foundation stones or any other signs of buildings, but you could tell where they once must have been—someone had planted four elm trees to provide shade for the settlers. Elms can live three hundred years when not struck down by Dutch elm disease. These elms have survived and now shade young stands of loblolly pines. As we stood on the bridge, the only sounds came from a bumblebee, the *conk-la-ree!* call of a red-winged black-bird, and a soft sigh of wind through the pines.

Oasis

People working on what may be the next greenline park in the United States say they were inspired by the Areas of Outstanding Natural Beauty in England. The 5-million-acre place they want to protect—in perpetuity, is the goal—is mostly privately owned and for four hundred years has been a mix of rich farmland, old forests, and almost impenetrable marshes. In the words of former Maryland congressman Wayne Gilchrest, who represented the area for eighteen years, it's "carpeted with farms, stitched with forests, and buttoned with rural fishing villages." Since AONB is too cumbersome a term, they're calling the project Delmarva Oasis. (Which is partly my doing. Back in 2017, in early meetings about the 5 million acres, I came up with OA—Outstanding Area—as a shortcut that could expand into "oasis.")

Nothing to do about the name "Delmarva," because since the 1700s this long peninsula, shaped like a teardrop with a tail, has been thought of as three different pieces. About half became Delaware (Del), while another not-quite-half is part of Maryland (Mar), and a little bit, the tail, went to Virginia (Va). It's never felt like any single place except that, because it runs along the Chesapeake Bay, it's been called the Eastern Shore. Ever since roads became prominent and the economy didn't depend on sailboats, the Delmarva Peninsula, which had been front and center in the Age of Sail, has been seen by the locals as at least "fifteen minutes too far away" to attract cities or industry.

So, a single place with a historically divided focus, which may be why no one has come up with a more user-friendly name.

Though "Delmarva" gets used occasionally: there are the Delmarva Shorebirds, a minor league baseball team. There's the Delmarva fox squirrel, twice the size of a gray squirrel, once endangered, now recovered, a shy animal with silvery fur and what's been called an exceedingly fluffy tail. And there are the Delmarva bays, seventeen thousand elliptical wetlands formed by strong winds sweeping over the peninsula during the Ice Age; each is about an acre in size and all have high concentrations of rare and endangered plants and animals. "If you could only save one type of wetland, this would be it," restoration ecologist Wayne Tyndall told Ashley Stimpson in *The Maryland Natural Resource* magazine. Delmarva bays were once considered "whale wallows"—caused (it was thought) by whales stranded by biblical floods.

"A sandbar" is how Wayne Gilchrest, the former congressman who now runs an environmental education center, describes the peninsula. A combination of silt and sand drained off the Appalachians to the west, and sand and sediment from the Atlantic to the east. The highest hill is only 101 feet tall. Driving around, the most common sight is a big field—there are nearly seven thousand farms, mostly growing corn and soybeans, feed for Delmarva's 605 million chickens. In the distance, a dense tree line, and arching above is the uninterrupted sky, a true horizon-to-horizon half dome. When the sun is low it casts exceptionally long shadows. As evening draws close, there are lingering sunsets that, Gilchrest told me, "pull people out of their homes with a coffee mug or a wineglass. You can't get tired of it."

In winter, the sky is an arena for tens of thousands of snow geese getting ready to fly back to the Arctic islands near Greenland, when the weather warms again. Cited as possibly the noisiest of all waterfowl, these large white birds with black wingtips sound like baying hounds when far away and, when close, are a

"chorus of shrill cries, hoarse honks, and high-pitched quacks," according to the Cornell Lab of Ornithology. Also in winter, down from the Arctic, are tundra swans. The wind whistles through their wings and they slap the water with their feet. Their sounds have been described as clarinet ensembles—reedy, piping voices.

Many of Delmarva's fields were laid out during prosperous days in the 1700s and have since remained intact, never giving way to urbanization or industrialization. Because of this, when you look at the peninsula today, "you can still see clear down to the eighteenth century and colonial America," A. Elizabeth Watson told me (she's the coauthor, with Samuel N. Stokes and Shelley S. Mastran, of *Saving America's Countryside*). It's a situation she describes as the rarest of the rare.

I'd never heard of the Battle of Caulk's Field, a turning point in the War of 1812, but gently sloping Caulk's Field (in the Mar part of Delmarva) is unchanged, and when the setting stays the same, long-ago events can seem close. Standing in the field one morning, I could trace the precise path British troops took on an August night in 1814 under a full moon, hoping their march would catch American militiamen unaware. Instead, only three Americans were wounded and fourteen English marines were killed, including their daring commander, twenty-eight-year-old Sir Peter Parker, a cousin of Lord Byron. One week before, the British had burned the White House. It's thought that news of the American victory at Caulk's Field inspired the defenders at Fort McHenry in Baltimore, where "the bombs bursting in air gave proof through the night that our flag was still there."

On Delmarva you can "see clear down" to America's shameful days as well as its triumphs. There was slavery throughout the peninsula until nearly the end of the Civil War. Harriet Tubman, born in the Mar part of Delmarva, escaped north to Philadelphia and freedom in 1849 when she was twenty-seven, traveling mostly at night: "When I found I had crossed that line, I looked at my hands to see if I was the same person." Over the next decade, as a conductor on the Underground Railroad, she made thirteen return trips to the peninsula to rescue seventy family members and friends, according to Kate Clifford Larson, author of *Bound for the Promised Land: Harriet Tubman, Portrait of an American Hero*.

During the Civil War, Tubman led an armed raid for the Union army that liberated more than seven hundred slaves from rice plantations in South Carolina. She was five feet tall, never learned to read or write, and relied on disguises and bribes. She seems to have had a remarkable geographic memory, following streams and constellations, and an unfailing sense of safety and danger. "I can say what most conductors can't say," she once told an audience. "I never ran my train off the track and I never lost a passenger."

When Tubman was twelve, maybe fourteen (no one's quite sure of the date, only the place), she was sent to the Bucktown Village Store to buy groceries—a one-room general-purpose shop with a porch out front. It's still standing and now operates as a museum. The same day I went to Caulk's Field, I stood on that porch and could see exactly where Tubman was standing when an overseer, trying to recapture a young slave who left work in the fields without permission, ordered her to help tie the slave up. She refused, and the young man broke free. The overseer picked up a two-pound weight and threw it to knock the fugitive down. His aim was bad, and the weight hit Tubman, fracturing her skull and bringing on severe headaches and seizures that would be with her the rest of her life. This was also the beginning of intense visions and dreams that she said showed her the future and gave her the skills to guide people north.

To Hell with Shell

Half of Delmarva is still in a natural state—there are places that were always too wet to farm, and others, now restored, where farming was tried and abandoned. The Great Cypress Swamp (mostly in Delaware, partly in Maryland) is a dense, dark, forested place where the light is always dim. Spotted turtles, globally threatened, are back, and so are carpenter frogs, brown and striped and two inches long, with a call that sounds like a hammer hitting a nail, sometimes two or three times, fast. Thirty years ago they were a rarity. But in 2015 ecologist Andrew Martin told *Delmarva Now*, "Last year I parked the truck and I thought there was heavy equipment moving back in the woods somewhere, and it was just thousands upon thousands of carpenter frogs."

The great natural spectacle in Delmarva is one of the great environmental phenomena on the planet: the meeting of two migrations every May, when red knots and horseshoe crabs return to Delaware Bay. Red knots—robin-sized sandpipers with a robin's red breast—winter in Tierra del Fuego, the southern tip of South America, and breed 9,300 miles away in the Arctic, many on Southampton Island, near the north end of Hudson Bay. Delaware Bay is their last stopover on the way north after several weeks of flying, sometimes without pausing for six to eight days at a stretch. To make it safely through the final two thousand miles, they need to double their body weight in about twelve days.

Which they can do because May is also when millions of tank-shaped, soccer-ball-sized horseshoe crabs, a species that's 450 million years old, return from out in the Atlantic to spawn on Delaware Bay beaches. Peak numbers arrive on the nights of a new moon and a full moon. Tens of thousands of red knots then gorge on the protein-rich eggs.

The spectacle isn't what it once was, now that researchers gather thousands of horseshoe crabs (their blood, which is blue, is used in medical tests), and it might have disappeared completely if not for Edmund H. "Ted" Harvey and Russell W. Peterson. Harvey, a hunting and fishing guide who founded Delaware Wild Lands, the state's first land trust, heard back in the 1960s that Shell Oil was planning to build a $200 million oil refinery on Delaware Bay. Looking at a map, Harvey started buying some of the land Shell would need, bits and pieces in a patchwork pattern, just enough to keep Shell from getting everything it wanted. He had bumper stickers and buttons made: TO HELL WITH SHELL.

What fully stopped Shell was that Peterson, a Republican and former DuPont scientist who helped develop Dacron and other "miracle fibers," became an environmentalist after he was elected governor of Delaware in 1968. He got the legislature to pass—by one vote—a Coastal Zone Act prohibiting heavy industry along the Bay and Atlantic shoreline for 115 miles. It was the first time, said Prince Bernhard of the Netherlands, founder and president of the World Wildlife Fund, that a community had put itself up against an international oil company and won. Summoned to Washington, Peterson was told by Richard Nixon's secretary of commerce, "Governor, you are being disloyal to your country." Peterson took to wearing a TO HELL WITH SHELL button.

"Futuristic Delmarva Oasis Plan Protects Delmarva Peninsula," said a 2018 headline on the Easton, Maryland, *Star Democrat* website. Rob Etgen, president of the Eastern Shore Land Conservancy, a land trust that partners with Delaware Wild Lands, said in the piece that the goal is to protect half the peninsula— its natural land and historic farmland—by 2030. This is right in sync with the Chesapeake Conservation Partnership, a coalition of more than fifty organizations looking to protect the larger Chesapeake Bay watershed, which spreads out over six states. Their slogan: "Half makes us whole."

Since 30 percent of Delmarva is already protected, as Etgen told me, getting to half sounds doable if it becomes a greenline park. Doable and necessary, he stressed, because by 2030 the peninsula may no longer be fifteen minutes too far away. Maryland has plans to add another bridge across Chesapeake Bay, and one of the alignments originally under consideration would have extended eight miles inland and run right over Caulk's Field.

A possible logo for the Delmarva Oasis

Looking beyond 2030, taking what Etgen calls a next-generation approach, will mean thinking about the seascape as well as the landscape. The Delmarva Restoration and Conservation Network, a group of government and nongovernment biologists brought together by the U.S. Fish and Wildlife Service, estimates that, in the next half century, as much as one-fifth of the peninsula will disappear under water. Sea-level rise due to global warming is being compounded by land-level fall, a kind of Ice Age seesaw called a "forebulge." Land to the north of the peninsula, pushed down by glaciers, is still rebounding, while the peninsula itself, which once swung upward to compensate, is still subsiding. The goal is to save 50 percent of the 80 percent that will still be above water.

It's not that Delmarva might become Delmar, though the Virginia tail is particularly low-lying. It's that sinking coastal salt marshes will need to retreat up to fifteen miles inland through what have been identified as "sea-level rise wetland migration corridors." Which can't happen if houses get in the way. So far, the most tangible signs of climate change are school buses that have been rerouted because of unexpected flooding, and "ghost forests"—stands of trees along the shore that have succumbed to saltwater intrusion. "It's dramatic," ecologist Greg Noe told the Associated Press, "and it's changing faster than it has before in human history."

Trying to protect land from getting cemented over at the same time it's melting away—a climate emergency and a Half Earth conundrum. Delmarva is a place where the "twin crises," as Jason Mark, editor of *Sierra* magazine, calls them, are inextricably intertwined. If it all works out, the peninsula will become a new kind of oasis for a warmer world, a greenline park created by its citizens. This could be the forerunner of a national system of twenty-first-century greenline reserves, finally filling in those dotted lines on the old treasure map.

A Planetary Perspective

I look forward with great optimism. I think that we
are experiencing not only an historical change, but a
planetary one as well. We live in a transition.

—VLADIMIR VERNADSKY

The 2020 bushfires in Australia killed more than a billion birds, mammals, and reptiles, and over 240 billion insects. More fires in the Amazon and other rainforests could, as *The New York Times* puts it, trigger a "self-destruct—a process of self-perpetuating deforestation known as dieback." Even at the height of the COVID-19 pandemic, harrowing warnings like these kept accumulating. Over the next several decades, what's it going to be for the planet—save or "6"?

While writing this book, I had the privilege to meet people around the continent who could be called extinction preventers, healers and rescuers, Half Earth practitioners like Steve Kallick, gardeners of a much, much larger landscape, the way Benton MacKaye was. Women and men Vernadsky looked forward to, people dedicated to strengthening and restoring regions of North America, thinking and acting with a planetary perspective.

This is some of their work.

For the exclusive use of animals

BANFF NATIONAL PARK, CANADA

Accompanied by Tony Clevenger, a road ecologist, I slithered down a steep embankment next to the Trans-Canada Highway, one of the longest highways in the world. After turning right, we picked our way through a dense stand of fir trees and turned right again, more cautiously this time, since we were approaching a point we weren't supposed to go beyond, then looked back toward the road.

It was a perfectly ordinary, utilitarian sight—a large, arched, corrugated steel culvert installed to let a stream, this one called Redearth Creek, pass under a highway; there are many thousands of these culverts beneath the nearly 5 million miles of roads, paved and unpaved, in North America. One detail was different: a slightly elevated, dirt-covered concrete walkway six feet wide along the side of the fast-flowing creek. Another difference: not many people get to see this culvert. Clevenger had to unlock a gate in an eight-foot-high, fifty-mile-long wire fence before we could get through, and he brought us to a halt well away from the bank of the stream.

The Redearth culvert is in the middle of Banff National Park, and we had reached what has become a biodiversity gateway. There are a series of them in the park, carefully designed, rigorously monitored for close to twenty years, and originally denounced as harebrained. In a voice just loud enough for me to hear, Clevenger, whose advice gets sought from Argentina to

China, told me, "We're looking at one of Canada's greatest conservation success stories. Twenty years ago, no one expected it to have far-reaching effects, and now it gets told time and again throughout the world."

Banff, the most visited national park on the continent, still has all the species it did when European settlers arrived in the nineteenth century. Another distinction: it's the only national park bisected by a transcontinental highway, the TCH, or Trans-Canada Highway, built in the 1950s and 1960s, like the U.S. Interstate Highway System. Nearly everyone who goes to Banff gets there on the TCH, which also carries trucks heading nonstop for the West Coast; it's an east-west thruway across what's essentially a north-south shaped park. A problem arrived with the TCH: WVC, or wildlife-vehicle collisions.

As on the interstates, drivers' needs are well taken care of on the TCH—the road is laid out so that at every moment they can see six hundred feet ahead. But, again as with the interstates, little thought was given to how having large animals nearby, such as elk and grizzly bears, would affect driving, or how the presence of the road would affect the animals. In the United States there are a million collisions a year between cars and large animals, and that number has jumped 50 percent since 2005, bringing about the death of 1 to 2 million large animals every year. And heavy costs to motorists who hit them, since repairing a car after colliding with a deer can average $8,000; with a moose, it's closer to $30,000.

Clevenger and I couldn't get any closer to the concrete walkway next to the culvert because it's constructed for the exclusive use of animals and they get spooked if people get too near. When the TCH got "twinned" in Banff, and the two-lane road became a four-lane divided highway, funds opened up for "highway wilding," as it's called. "In the summer," Clevenger said, "thirty thousand vehicles a day move past Redearth Creek. That's a car every three seconds. But for the animals it's as if the highway isn't even there."

Hiding the highway from the animals meant tunneling for lots of culverts with walkways, putting up miles of fencing so the animals can't dart across the road, and building half a dozen dirt-and-tree-covered overpasses. "Landscape bridges" is Clevenger's name for these wider, open-air crossings. People aren't supposed to notice them, and there are no signs announcing what people are passing under, tempting them to get out of their cars and wander into the animals' right-of-way. Like the culverts, the overpasses are nothing much to look at—everything's made from standardized parts, and boulders are grouped in front to conceal some of the concrete.

But then, that's not the point. Andrew Evans of National Geographic, one of the few journalists who got to walk across an overpass, saw the tracks of black bears, grizzly bears, wolves, elk, and deer, and wrote that he was "amazed how natural the area seemed. The noise disappeared, as did the road. The forest has filled in the area so that anyone crossing would really have no idea they were crossing a major highway."

There are animal cams at all the wildlife crossings, and patterns have emerged—grizzlies, elk, and moose like high, wide, open crossings; black bears and mountain lions prefer long, dim, narrow tunnels. Elk start using an overpass before it's even finished. Grizzlies and wolves take up to five years before they feel comfortable, though once bears bring cubs across a bridge, or mountain lions bring kittens, the next generation crosses without hesitation.

Just as important, WVCs are down 80 percent, and since 1997 there have been nearly two hundred thousand safe animal crossings over and under the TCH by all eleven of the park's large predator and prey species. Even by wolverines, so fierce and wary they've been called a thousand pounds of attitude in a

Wildlife overpass, Banff National Park

thirty-pound body. Beavers, toads, and garter snakes also use the crossings.

Wildlife crossings aren't new—it's been thought they got started in France in the 1950s, as an aid to hunters chasing deer. Paul Drummond, a young landscape architect, has traced them back to the late 1920s, when William du Pont Jr., who bred race-horses and designed racetracks, built overpasses and underpasses on his seven-thousand-plus-acre estate in northern Maryland. That way he and his friends could gallop unimpeded across the countryside on foxhunts and steeplechases. Drummond told me he got permission to set up his own animal cams on du Pont's estate, now a state-owned natural resources management area, and found that, close to a century later, the culverts still act as wildlife crossings, attracting deer, rabbits, raccoons—and an occasional fox.

Photos taken by Clevenger and his team at Redearth Creek

show a mule deer, a grizzly, a cougar, and a wolf using that underpass—snapshots of a reconnected world. Even from where I stood, I could see deer tracks. It was quiet down there, no traffic whoosh, the loudest sound the splash of the stream. By 2030, there will be 2 billion vehicles in the world. By 2050, 15.5 million more miles of road. Clevenger thinks it's time to make "roadkill" an obsolete word. He mentioned wildlife crossings being created in Washington State, in Montana, Wyoming, Nevada, California, and Florida. "If you build a network of these bridges," one of his colleagues, ecologist Nina-Marie Lister, said in a *Vox* interview, "you've solved the problem for good. It's done."

Holdfast zones

THE BERKSHIRES, MASSACHUSETTS

Sitting on some large flat rocks at the base of a sixty-foot-tall, plunging, tumbling, cascading waterfall on Sanderson Brook in an old state forest in western Massachusetts, three of us unwrapped sandwiches. There was no one else around, and the waterfall is a doozy (in the words of one blogger). Within a mile is a trail to the top of a steep-sided hill with a vista overlooking hundreds of thousands of acres of nearly continuous forests.

Not many people find their way here, to Chester-Blandford State Forest, which is shared by and named for two hill towns in the "undiscovered" southern Berkshires. Both have smaller populations now than they did in 1850. The state bought the land in the 1920s, from timber companies that logged it and left. It would be easy—much too easy, it turns out—to think of this as a place that early on was picked clean and since then hasn't had much to offer.

Small, played-out mine shafts from the 1800s dot the area—

there's Gold Mine Brook and Mica Mine Road within Chester-Blandford State Forest. But there's also Beulah Land Road, a reference to Isaiah's biblical prophecy that the Hebrew people will find themselves and Israel transformed after they return from the Babylonian captivity: "Thou shalt no longer be termed Forsaken; neither shall thy land any more be termed Desolate; but thou shall be called Hephzibah and thy land Beulah."

This is not yet an old-growth forest, more like a middle-aged one, since logging only ceased about a century ago. The wide path to the waterfall was cleared in the 1930s by a Civilian Conservation Corps crew, kids from Boston for the most part. Unfrequented by people, the woods have gained a population of moose, black bears, bobcats, songbirds, salamanders, and dragonflies.

Andy Finton, the landscape conservation director for The Nature Conservancy in Massachusetts, one of my companions, sees the Beulah-land qualities of this forest: as he shows me on a sheaf of maps, Chester-Blandford has a rare and exceptional and lasting ability to hang on to its plants and animals. The assumption, common among ecologists, has been that as the earth warms, species can only survive by moving north. But here's a holdfast zone, a stronghold of life, a bulwark against the disappearance of biodiversity in the face of climate change.

This idea springs from resilience science, a new discipline that has identified a network of more than one thousand such places in the East, extending across 101 million acres, or 23 percent of the landscape. In the restrained language used by resilience biologists, this is land that's "far above average"—their highest praise. What they've found is a reliable correlation between the enduring physical configuration of a place and the life it shelters.

More specifically, topography is an indication of biological persistence. Resilience maps record the presence or absence of seventeen different landforms, like "steep slope cool aspect," "steep slope warm aspect," "summit/ridgetop," "wetlands," and "valley/toeslope." Chester-Blandford has most of the list. The sharp plunge of Sanderson Brook Falls makes a steep slope unmistakable, but other distinctions are less easy to spot. Given the uniform blanket of green that covers eastern forests, even a precipitous slope can, for instance, look like nothing more than a gentle roll or rumple.

How does this resilience work? Varied topography creates tiny, self-contained microclimates that, when crowded together, put a range of hot, cool, wet, dry, windswept, and sheltered alternatives within easy reach. No matter what happens to the regional climate, resident species will have options, "whether," as Finton says, "you're a beetle or a turtle or even a seed or a spore."

"Or put it this way," says Mark Anderson, the director of conservation science for The Nature Conservancy's eastern U.S. region and the third member of our party. "There are no microclimates in the middle of a large parking lot. But in a nearby natural area on a blistering hot day, you can get out of the sun or the wind and duck into some shade and find water." So, for plants or animals, staying put in a resilient site actually means making tiny migrations into new microclimates right next door. Wiggle room confers stability.

"I really didn't see this coming," Anderson says about the way his resilience work has grown. He has a trim goatee and a ready smile and is in constant demand around the country. He possesses "the chair," a wooden armchair given to anyone who's worked for the Conservancy for twenty-five years. The past dozen years have been particularly intense. He's been leading groups of biologists and ecologists who are mapping resilient landscapes across the United States and part of southern Canada. The rest of Canada is talked about, and Mexico is hoped-for.

The project has been jump-started by the fact that much of

North America is "premapped"—accurate topographical and geological maps, now digitized, became available shortly after the Civil War. Knowing the contours of the countryside, Anderson and his researchers could pinpoint and color-code potential sites even before investigating them. Before this trip, Anderson himself had never been to Chester-Blandford, despite its being shaded blue on Finton's maps (the "far above average" color). The maps also display connectivity, showing how well linked the forest is to other nearby resilient sites, which means the forest could be a path for a northward flow of species.

Anderson's work is resolving a back-of-the-mind Nature Conservancy concern—are they protecting the right landscapes, the places where biodiversity will flourish in the twenty-second century? By and large, yes. More challenging is the finding that, in the eastern United States, only 44 percent of one thousand resilient sites are protected. This has inspired the biggest project in the Conservancy's history, a pledge to protect another 25 percent of these sites nationally by 2050. Since they're spread out over 23 percent of the East, and over 24 percent of the Great Plains and Great Lakes, does this make Anderson a Quarter Earther? "Well," he says, "if Half Earth is the goal, here's the quarter to start with."

A crown of creation

PAINT ROCK FOREST, ALABAMA

I was on a Kawasaki MULE 3010, an ingenious cross between a golf cart and a mountain goat. The driver, Bill Finch, paused the machine at a precarious angle and began pointing out different species of oak trees, counting eleven within a couple of minutes. Bill Finch, whom E. O. Wilson has praised as the best all-around naturalist he's ever known. The MULE, an acronym for "multi-use light equipment," is essential for scampering around the kind of terrain in northeast Alabama that Finch was showing me through: a series of steep, forested slopes alternating with flat places on one side of the Paint Rock Valley, which slices across the Cumberland Plateau—rugged, remote hillsides on the west side of the Appalachians.

Only recently has Paint Rock Forest, which is perhaps a million years old, been called one of the crowns of creation. It's the very center of North America's hardwood or deciduous tree diversity, meaning trees that shed their leaves. Which is why Finch, along with the University of Alabama and Alabama A&M University, a historically black college in nearby Huntsville, is busy implementing North America's most intensive forest research program; it will be complete in 2070. I was there for the program's inauguration, in the early spring of 2018.

Finch was wearing a floppy sun hat, as he always seems to be, and has a graying full beard and curling mustache. He studied forestry and liberal arts in college before becoming a newspaperman and director of the Mobile Botanical Gardens, and he's the host of a Sunday-morning gardening program on what the rest of the week is Alabama's most conservative talk-radio station. Finch first noticed Paint Rock in the late 1970s, while driving home from college for Thanksgiving through the Paint Rock Valley, which itself has kept a Brigadoon-like quality—it's sometimes called the little sister to Virginia's Shenandoah. Supposedly Davy Crockett carved his name on one of the trees there, and many houses didn't have electricity until after World War II.

What caught Finch's eye was the "rainbow effect" in the autumn leaves up the slopes, gold and red and "this weird claret and an orangey, pinkish purple." Such an intense burst of color coming from so many different kinds of trees that he stopped his truck in the middle of the road, and stared. "Well, I had to," he told me. "It just slapped me in the face." Even though this contradicted much of what was known about the landscape—namely that however pretty the area was, it wasn't supposed to be all that ecologically interesting or diverse or remarkable.

Tramping through Alabama woods as a boy, Finch once saw a red buckeye, a small tree called the firecracker plant, in full bloom, its brilliant, dark red flowers in tubular clusters—"like the flaming sword at the gates of paradise," Finch says, the one preventing Adam from reentering the Garden of Eden. When Finch got home, no one could identify the tree, to his deep disappointment. It was the first time he guessed there was something askew when it came to people's knowledge about what was out there in the landscape.

Later, he looked for books about Alabama's forests, and back then found only one, written in the 1890s, by Charles Sprague Sargent, the Waverly Oaks arborist—Sargent was a quartermaster for the Union army in the Civil War, during the Battle of Mobile Bay. "Essentially it's taken us five hundred years to get to our research program," Finch told me on the MULE. "In too many ways, there are too many Americans who just got off the boat with Columbus."

Meaning . . . ? I asked.

"European settlers," he said, "were unprepared in terms of language, expectations, and perceptions for what they would encounter in the New World. Northern Europe has four species of oaks, so they brought very few oak names with them. That's why we still only talk about red oaks and white oaks, although Alabama has twenty-three kinds of red oaks and seventeen kinds of white oaks. Europeans have four seasons, but Mobile has six. Australia actually has eight. The only new name we have is New England's mud season between winter and spring."

Finch said that by ignoring the Native American presence in the Cumberlands, scientists continue to ignore seventeen thousand years of experience about life on the continent, its ecosystems, and the seasons. Even naturalists known for their thoroughness ignore this side of the Appalachians. Asa Gray, a friend of Darwin's, a great scholar, and considered the father of American botany, chose to spend many summers on the east side of the central Appalachians, in Virginia and North Carolina.

"It was assumed for a hundred and fifty years," Finch said, "that Gray had found the heart of forest diversity, although you and I just saw as many oak species as there are in the entire Great Smoky Mountains National Park, which covers half a million acres in North Carolina and Tennessee. The eastern slopes of the mountains are cooler, not as hot—so that's where the trains went. I have a theory, only a theory, that Mrs. Gray didn't want to spend Julys in Georgia or Alabama, and that's why our understanding stopped several hundred miles short of the truth."

For years conservationists ignored the Cumberland Plateau. "It was seen as the working side of the Appalachians," Finch said, "no place for tourists, only good for mining coal." *Night Comes to the Cumberlands: A Biography of a Depressed Area,* a 1962 book by Harry M. Caudill that quickly went through twenty-two print-ings, focused on eastern Kentucky and inspired Lyndon Johnson's 1964 War on Poverty. It brought to national attention this story of disfigured land and scarred lives, calling coal a "crown of sorrow." Caudill said, "Coal always cursed the land in which it lies."

Caudill proposed a billion-dollar federal program to create jobs, schools, and housing, and also wanted to see half of the Cumberlands set aside as protected forests (this idea was flatly ignored). What Caudill didn't spell out was how huge the areas are that have *not* been devastated—maybe 10 to 20 million acres from Kentucky to Alabama, with Paint Rock at the core.

Finch had a lot to show off as we MULEd past drop-offs with a seventy-degree slope. "For a while, we knew a good deal about the rare species here," he told me. "Some of them spectacular, like yellowwood, with long, drooping, wisteria-like sprays of white flowers that for some reason are at their best every other year. Or Alabama snow wreath, whose small white clusters of flowers look like clumps of just-landed snow. Or the American smoke tree—its feathery blossoms extend from the ends of branches like puffs of pale lavender mist, or smoke. The wood of this tree is unique. When the tree falls, it lies unrotting on the ground for up to a century. So many kinds of trees thought to have been lost to the world manage to survive here. Why is that? We're trying to find out."

A month later at Paint Rock, Finch discovered a wild azalea, a single stand of bushes where the flowers are all different colors— pink, pink with white stripes, peach, and yellow with peach splotches. "It sounds like a modern hybrid," he told me, "but it's all by itself, and I suspect these azaleas are an ancient population whose genes got exchanged tens of thousands of years ago as they adapted to attract lots of different pollinators. Very fragrant, also very funky."

Our MULE was taking a zigzag path. Like many Cumberland Plateau hillsides, this is a forest interlaced with caves. Plunging streams appear out of nowhere and disappear into a hole in the ground. Go down into some of the caves with lights, and you can find rivers with beaches, pools, shoals, and waterfalls.

We were passing through what Finch calls the most extravagant display of spring ephemeral wildflowers in eastern North America—such as trout lilies, Dutchman's breeches, trilliums. These wildflowers show up in undisturbed woodlands after snowmelt and are gone by the time trees leaf out, having already produced seeds that ants carry across the forest floor. Because an ant can move a seed only so far, if the flower's habitat is disturbed it quickly disappears. But at Paint Rock they are everywhere.

Our destination was Calloway Sinks, a spot at the top of the forest where the land slightly collapses in on itself, creating bowl-like depressions that had become a bluebell wood. The place seemed stark, empty, and colorless—except when you looked down. Every bit of the forest floor was covered by an inch-high tapestry of soft green leaves supporting millions of shimmering miniature light-blue flowers on tiny, slender stalks.

We were in the middle of acre after acre of wildflowers—fields, meadows, and pastures—and still surrounded by bare tree trunks. There are bluebell woods on other continents, and one hypothesis says those are the remnants of a global forest, the Arcto-Tertiary, from millions of years ago. Which might explain the from-another-time feeling pervading Calloway Sinks.

At one point the MULE stopped to offer a lift to Doug Booher, a young myrmecologist, or ant specialist. Booher grew up with a love of walking through the woods, discovering things, and then ran a company making floors and furniture out of timber salvaged from old buildings. He got a PhD in ecology so he could get back outside again.

When we caught up with him, Booher, who has a showy, cascading mustache, was spending several days prowling around Paint Rock, turning over rocks and leaf litter to help Finch get a better idea of the animals on the property. During one ninety-minute trek, Booher found three ant species unknown in North America, along with a beetle and a fourth ant species, neither of which had ever been seen anywhere.

"Did you ever notice," Booher asked us, "how when you're down close enough, the soils in Paint Rock have sharply different smells? Some are grassy and green, like an open meadow. Some have a fermented, mushroomy smell. I've pretty much had a smile on my face the whole time I've been here." Other naturalists have reported a high number of salamander species, an unexpectedly huge concentration of land snails (reported as disappearing at an astronomical rate elsewhere), and a diversity of cave-living animals unmatched anywhere except China.

Finch plans to count and name all the trees at Paint Rock. For forty years there's been a rigorous way to do that, a technique pioneered in a Panama tropical forest, Barro Colorado Island, focusing on sample plots of thirty by thirty feet and repeated every five years. What's counted is every tree that at breast height is at least as thick as a pencil.

Finch has a hunch his censuses will explain what he calls the twin foundations of the place—persistence and patchiness. Many northern forest species retreated down to the Cumberlands during glacial times, but some trees didn't altogether abandon the area when the ice ages ended. Ever since, the varieties that left individuals behind have been able to tolerate Alabama's July and August heat. That's why there are so many places at Paint Rock where Finch, standing still and in a single glance, can not only see the oaks he knows so well, but six or seven different types of hickory trees, four species of elms, and four or five maples and ash trees. At the same time, there's streakiness—some of the rarest species, like American smoke trees, grow only in strips about 150 feet wide.

"What I suspect," Finch said to me, "is the answer to these anomalies will be found underground, in Paint Rock's soil, the mix of minerals, water, oxygen, and tens of thousands of microorganisms—the part of the forest we know least about and have to get to the bottom of, literally. We know soil can either nourish or discourage plant life. Clearly the soil here is hospitable to the rarities here, helping them develop extra genetic material to withstand the heat and fight off disease. Now the soil will get its own census. I think we'll see it's every bit as complicated as what's aboveground. Only more so."

The water gentler

LOS OJOS RANCH, SONORA, MEXICO

Just south of Arizona, on an old cattle ranch no more than two miles below the U.S.-Mexico border, the landscape is curiously striped. One part is brown, austere hillsides, dry, stark, and rugged. But spilling down the hills in widening bands, and spreading across valleys, are vivid splashes of green. One morning, in the biggest green patch, I took a walk with the ranch's owner, Anna Valer Clark, a painter from the East Coast who bought the place thirty years ago. There were only two sounds to be heard—the murmur and gentle burble of a shallow river at our feet, and the soughing of the wind, sometimes intense, sometimes fading away, swaying through the leaves of the trees high overhead.

The cattle at Cuenca los Ojos are long gone, but even on a short walk along the river, you can catch sight of a black bear, a roadrunner, a Gila monster, the whittled stump of a willow tree that had been gnawed on by a beaver, and a Gould's turkey, the largest of wild turkeys, whose dark feathers glow with a greenish, coppery iridescence.

What I couldn't see were all the piled-up stones at Los Ojos, thousands and thousands of them. These stones are what's turned the brown land green, providing shelter and sustenance for animals that had vanished and have now returned. Clark, who is called by her middle name, Valer, refers to these carefully placed rocks, most of them hidden underground, as "my work."

Clark is a wilderness architect, a water gentler, someone who moves rocks around to change the way water flows. Her discovery was accidental, she says, worked out step-by-step and slowly over decades. When rainstorms arrive in an arid region, the rocks, if arrayed and deployed across the landscape, can act as a kind of flash-flood extinguisher. Reconfigured rocks keep water in place, turning it from an occasional torrent that sluices and slices through narrow and ever-deeper arroyos (mini-canyons) into a spreading trickle that sticks around, allowing seeds to sprout and grasses and trees to take root. "Four geological forces reshape the earth," Ron Pulliam, a friend of Clark's and former president of the Ecological Society of America, told me. "Vulcanism, tectonic plates, erosion—and Valer."

It was Clark's chance encounter with the *trincheristas*— the terrace builders—that changed the course of her life, and the landscape. *"Trinchera"* is a military term, referring to the trenches soldiers dig to protect themselves from enemy fire, and also a farmer's term for walls thrown up to enclose hillside ter-

Anna Valer Clark

races, creating bands of flat land on steep slopes wide enough for growing plots of chilis, beans, and corn.

In 1983 Clark, whose father was a Wall Street broker, bought her first ranch, a dusty, hilly, once-forested, overgrazed, and underwatered cattle ranch in southeast Arizona, thinking it might be a vacation home. Seeing bare, drought-stricken hills, and scarcely any grass, she thought, "I've got to do something about this." Now she says, "I wasn't thinking about rearranging the planet. I just wanted to improve the land I was standing on." Her method—small, practical steps, and common sense.

Overgrazing could be handled easily enough—a matter of shifting cows around often enough so they never got a second bite of the same plant. What about the water? There was one spot on the bone-dry ranch where too much water was actually a problem, sloshing across a road. Migrant Mexican ranch hands, whose families lived in hillside villages hundreds of miles south, had come north for harvesting jobs and fixing fences. Clark asked some of these workers to throw down a few rocks across the spillover.

Immediately it was as if the water had found new purpose. It slowed, no longer slopping over the road; instead, back behind the rocks, it pooled and started seeping into the ground. A year later, in 1984, the ground was still damp and green shoots had pushed up. In southern Mexico, farming had been impossible without *trincheras,* and it just so happened the ranch workers were prodigious stoneworkers. Craftsmen. They could make watertight walls simply by choosing rocks that fit together seamlessly and without mortar, something they learned from their fathers and grandfathers. Maybe this exacting geometric farming skill could be applied to conservation?

There were certainly more than enough rocks—"Well, basically it was all I had to work with," Clark told me. "So I picked up a rock." You could start at the top of each hill and, moving downslope, run parallel lines of small rock dams across every dry stream bed, so that when the rains returned—and there were violent thunderstorms every July, locally referred to as monsoons— the water would be descending a staircase, not a slide. It was only years later, she said, that scientists told her what she was doing was unique.

There are now forty thousand *trincheras* on the eleven ranches she has owned. These are supplemented by fifty gabions, constructed of big wire baskets of rocks that get tied together and laid down in layers, another old military technique. Clark's gabions are large stone dams up to a thousand feet wide. Research by the Universidad de Sonora reports huge changes in her property as a result, including the reappearance of eighty-five bird species and twenty-four kinds of mammals.

Also the disappearance of smugglers—one dry riverbed was so deeply incised into the ground that trucks making for the border could drive through it undetected. With gabions in place, soil has accumulated and the ground is twenty feet higher, not a hiding place. The rain harvesting, made possible by stone speed bumps, has nutrient harvesting as another benefit, because the dams also hold on to any topsoil or silt or fungi that the more slowly moving streams have picked up. By one estimate, desert soil ordinarily has 0.5 percent organic matter, but the green land like that on Clark's ranches can have 5 to 10 percent organic material—what you'd expect to find on a fertilized midwestern cornfield where tallgrass prairie once grew. The piling up of the soil has largely buried the *trincheras* and gabions, as if all that were visible of a cathedral was a roof, and the arches and flying buttresses had become underground infrastructure.

On Clark's ranches, rivers and streams flow year-round and naturally occurring springs have reappeared. Ichthyologists (fish experts) persuaded Clark to buy Los Ojos because it had a stream with eight species of fish that might be lost. "Los Ojos" means "the eyes" in Spanish. In Mexico, it also refers, poetically, to "eyes in the ground"—springs, small groundwater-fed oases and wetlands, or *ciénegas*. Ninety-five percent of *ciénegas* vanished over the past two hundred years during what's known as "the Great *Ciénega* Disappearance." To Native Americans like the Hopi, says conservationist Craig Childs in *The Secret Knowledge of Water*, *ciénegas* were "simply alive. They were points where creation came to the surface and spilled out."

The multinational landscape where Clark works is sometimes called the Apache Highlands, and also the Sky Islands—even before people arrived, it was a crossroads region of wildlife profusion because of what pours into it from all sides. "If you had to pick one place in the entire continent," William deBuys, a conservation historian, writes in *A Great Aridness*, "where the greatest number of surprising plants and animals mingle in proximity to each other, you would do well to stick your pin in this part of the map." There are a dozen kinds of hummingbirds, and one mountain has more than half the bird species found anywhere in the United States.

"Less well known," writes Arizona conservationist Matt Skroch in an online essay, "is the fact that the Sky Islands also host the greatest number of mammals in the U.S."—more than twice as many as Yellowstone, including jaguars and ocelots. Plus three thousand plants, two hundred butterfly species, and up to eight hundred kinds of bees that nest on the ground and on twigs and emerge after intense bursts of rain. Ecologist and writer Gary Paul Nabhan says, "It's *the* place for bees."

To the north, reaching through Canada and into Alaska, are the Rockies and beyond them the Brooks Range—mountains for three thousand miles. To the south the mountains become, for another thousand miles or so, the Sierra Madre Occidental. To the west is the Sonoran Desert with its forests of towering saguaro cactus, and to the east, the higher elevation Chihuahuan Desert, considered the most biologically diverse desert in the Western Hemisphere.

In the midst of it all, temperate climates meet subtropical ones. This overlapping contributes to the "eruption of life," in Skroch's phrase. Conservationists also see the area as the link in a future tri-national system of protected land along the continental backbone. As Clark told me, "Eventually the three countries will hold hands." Ron Pulliam remembers how in the 1970s a colleague of his, Peter Warshall, a biologist considered a "biogladiator" for his outspokenness, sketched on a napkin routes through the Sky Islands that birds, large cats, and insects could use to travel from the Sierra Madre to the Rockies.

A contemporary Chicana poet and writer, Norma Elia Cantú, has written about the conflicts among the myriad groups of people who settled in or moved through the border region: "The pain and joy of the borderlands—perhaps no greater or lesser than the emotions stirred by living anywhere contradictions abound, cultures clash and meld, and life is lived on an edge—come from a wound that will not heal and yet is forever healing. These lands have always been here; the river of people has flowed for centuries."

Some currents from the river of people have only left traces; others are vividly remembered. SHOOTOUTS DAILY proclaim signs in Tombstone, Arizona, aimed at tourists who come to see reenactments of the Gunfight at the O.K. Corral; loudspeakers play the theme from *The Good, the Bad, and the Ugly*. At its most prosperous, the town, where silver miners vied for priority with ranchers, had 4 churches and 110 saloons, but the boom time lasted less than a decade.

For a while there were Mormon colonies south of the border (Mitt Romney's father was born in one), but many were evacuated over a century ago during the Mexican Revolution. Vanished U.S. prosperity is on display in the lobby of the Gadsden Hotel in the border town of Douglas, Arizona, built in 1907 out of profits from the 3 C's—copper, cotton, and cattle. There's a forty-two-foot-long Tiffany-style stained-glass mural of saguaro cactuses under a mottled sky; the columns are adorned with gold leaf; and there's a chip on the seventh step of the Italian marble grand staircase, supposedly knocked loose when the flamboyant Mexican revolutionary general Pancho Villa rode up the steps on his horse, "Siete Leguas" (Seven Leagues).

In 1886, the Apache Wars formally ended when Apache leader Geronimo, after more than thirty years of fighting and by then fifty-seven, surrendered after pursuit by a quarter of the U.S. Army. Avoiding captivity, a group of his followers took refuge in the northern Sierra Madre, where for decades rumors of their presence led miners, loggers, hunters, and ranchers to look elsewhere for new opportunities. Traveling to these mountains in 1937, conservationist Aldo Leopold marveled at the unspoiled wilderness with its full array of animals and plants. For him, historian Richard L. Knight said, this was "what an ecosystem resembled on the 'eighth day of creation.'"

Not much escaped the loggers who found the area after World War II, but there's a two-hundred-mile-long stretch of the Sierra Madre that highways have yet to cross. Gary Nabhan points out that Leopold seems to have come into contact with a sculptured hill that may have been left behind by precursors of Clark's *trincheristas*—the Cerro de Trincheras (Hill of Terraces), a five-hundred-foot-tall volcanic mound in Sonora ringed from bottom to top by nine hundred terraces at least five hundred years old.

There's a strong population of Mennonite farmers in the region, blond and blue-eyed and originally from Holland. Centuries ago they made stops on the plains of Ukraine and western Canada; they speak Plautdietsch, a unique Flatland German (the numbers begin *eent, twee, dree*). Their farms look right out of Illinois or Iowa, since, rather than capturing and cajoling intermittent rainwater, many Mennonite communities drill deep wells and convert desert grasslands into giant irrigated fields of soybeans, corn, and cotton.

Nearby, on Rancho El Uno, a conservation ranch run by a Mexican foundation, a herd of two hundred bison, the southernmost herd on the continent, is perceived differently on either side of the border. In Mexico bison are considered an endangered species; in New Mexico they're classified as livestock and can be hunted. Long before President Trump suggested building an impenetrable eighteen-foot (or higher) wall along the border, one that even low-flying pygmy owls couldn't surmount, Interstate 10, which runs through Arizona fifty and more miles to the north (it was completed in the early 1960s) separated animal populations that had previously freely moved north and south. Interstate 10 has yet to be retrofitted with wildlife overpasses and tunnels.

Gary Nabhan calls Clark a countertrend. Standing in the middle of a caldera, a large extinct volcano that was once an enormous wetland and was then systematically drained, Clark, ramrod erect with shoulder-length white hair under a big straw hat, looks forward to the work that lies ahead. Los Ojos Ranch is only half greened, and wetlands now cover 15 percent of the caldera; they had been reduced to 4 percent. She's created a nonprofit organization in Mexico and the United States, Cuenca Los Ojos (Bowl of Springs), to take on permanent ownership of her lands, and she thinks of herself as a partner to an ever-growing number of people and groups strengthening the Apache Highlands. Like Ron Pulliam, who has been able to turn a 1,200-acre bankrupt subdivision near Patagonia, Arizona, into a wildlife preserve.

"Some days we can't believe we exist," Clark says. "But it *is* a movement." Made up of people and groups whose work, like *trincheras* and gabions, is mostly unseen by those elsewhere in North America. Or appreciated. Pulliam remembers flying to Washington, D.C., with Clark so she could accept an award from a federal agency for restoring streams on one of her ranches. Meanwhile, another federal agency was suing her for illegally restoring a stream in a national forest next to that ranch. (The suit was later dropped.)

"I've been asked, 'Why don't you buy the best land and preserve it?'" Clark says. "I always reply no, I buy the worst land and turn it around. It's not just that if I can do it, anyone can—it's that if it can happen here, it can happen anywhere."

The Piney Woods

FLORIDA PANHANDLE

A few years ago, E. O. Wilson took me on a trip to meet M. C. Davis, a friend of his who was growing a forest in northwest Florida. Davis, a multimillionaire commodities trader, grew up in a Florida Panhandle trailer and raised his first stake playing poker. Like Wilson, Davis was tireless and an elaborately courteous southern charmer, but there was, Wilson said on the way, one big difference between them: "I only write about saving biodiversity. He's actually doing it."

Davis's idea was to revive the Piney Woods, the signature ecosystem of the American Southeast. A longleaf pine forest once covered 90 million acres, or about 60 percent of a virtually continuous 1,200-mile stretch across nine states from Virginia to Texas. That forest has been reduced by 97 percent, and there are only about 3 million acres left. Which is more catastrophic than what's happened to the Amazon rainforest (over 20 percent lost), or to coral reefs (30 to 50 percent destroyed). The longleaf pine forest's Big Cut, as it's still known, began after the Civil War and left behind what's referred to as "a sea of stumps." Much of the land has been reforested, but de-longleafed, and is now planted with row after row of faster-growing pines raised for pulpwood.

In 2000, Davis, who was a folksy, slightly rumpled, and unassuming man—"I'm a dirt-road, Panhandle guy," he told me—began spending half a million dollars a year planting longleaf pine trees, and another half million on other parts of a longleaf forest. He remembered the awakening that started it all.

He'd gotten stuck in a big pileup on I-4 near Tampa, saw a high-school marquee with the sign BLACK BEAR SEMINAR, and walked in the door: "There was an old drunk, and a politician who'd thought there'd be a crowd, and a couple of Canadians looking for day-old doughnuts and coffee—and, up on the stage, two women were talking about saving black bears. They were riveting. The next day I gave those ladies enough money to keep going for another two years—which I think scared them, it was so out of the blue. Then I asked them for a hundred-book environmental reading list for me, for my education. I spent a year reading Thoreau, John Muir, Ed Wilson. Then I started buying land to see what I could do."

If you were going to save Florida black bears, it was clear, you'd have to save longleaf forests, one of their habitats. An adult male black bear roams across perhaps 120 square miles of land. North Florida already had some good-sized clusters of publicly owned longleaf—national forests, state forests, wildlife management areas, and in the western Panhandle, Eglin Air Force Base, a huge facility that back before World War II had itself been a national forest. If you could add in enough territory to put these pieces together, they'd amount to something greater than a "postage stamp" of the natural world, as conservationists were referring to national parks. The problem was that 70 miles separated the first two protected longleaf forests—and it was another 95 miles to the third.

As Davis dug deeper, he realized the coastal Southeast is a hyperdiverse biological hotspot with up to sixty species in a single square yard—though you might not think so to see it, since a mature longleaf forest looks clipped and kempt, more like a city park. Without any human intervention, here is a forest with tall, straight trees that are rather widely spaced, letting in plenty of sunlight, and there are lots of open, grassy meadows. A longleaf tree only branches out once it's high overhead, where glistening needles up to a foot and a half long are arrayed in pompom-like sprays. Below the branches is empty space a hawk can glide through.

Davis's plan was to buy up and re-longleaf the "in-between" open space east of Eglin and west of a protected corridor along the Choctawhatchee River. The land was close to people, just a few miles inland from the sugar-white sands and high-rise condos of Gulf Coast beach towns. These tourist-driven communities used to be known as the Redneck Riviera, featuring attractions like the Snake-a-Torium, but more recently have been marketing themselves as the Emerald Coast (with slightly confusing slogans like "White Sand, White Wine, White Necks"). There was nothing, however, even remotely upscale about the land Davis had his eye on. It was dismal rather than dazzling, a series of abandoned peanut farms and unproductive pulpwood forests with lower asking prices.

The approach was called "M.C.'s folly" by conservationists because it seemed too ambitious. But Davis persisted. "Ed set the course," Davis told me, "by showing us that doing something huge is our only hope. We're all marching under his umbrella."

Davis bought 51,000 acres of degraded farms and forests, a piece of land up to twelve miles wide that included barely 1,500 acres of longleaf pine in scattered patches. Basically, he'd be starting from scratch. Davis named his bedraggled purchase Nokuse Plantation. Pronounced *No-GO-see*, Nokuse means "bear" in the language of the Muskogee people who once lived there, but their written alphabet doesn't have a hard G. Nokuse is the biggest private preserve and the biggest restoration project east of the Mississippi.

Then, to honor Wilson, Davis built the stunning $12 million E. O. Wilson Biophilia Center at one edge of Nokuse, where thousands of fourth through seventh graders from six counties get free classes that let them hold baby gopher tortoises and clamber and pose for pictures on a giant ant sculpture.

Even after years of work, much of Nokuse is still scruffy. A longleaf reforestation turns out to look like a construction zone, as Davis acknowledged while driving over bumpy trails in a golf cart. "Well," he said, "I tell people we're now in year thirteen of a three-hundred-year program. I could easily make a thousand acres look beautiful, but the extinction clock's ticking, so I decided to take on the bigger challenge."

At Nokuse, Davis and his crew had already thinned 22,000 acres of pulpwood pines and planted 8 million longleaf seedlings on 24,000 acres. He brought flames back to the woods after a half-century absence, setting carefully controlled fires on about 10,000 acres every year. For the past 25 million years, a prominent feature of the weather in this coastal environment has been violent summer thunderstorms and strobe-like lightning strikes.

What grew here, uniquely, was a fire-and-rain forest, one that, to stay healthy and keep its open glades, thirsts as much for scorching as it does for drenching—the one starts seeds germinating, the other lets them grow. Longleaf itself only thrives because it has evolved a slow, intricate fire dance that lets it evade being burned. An infant longleaf looks like a clump of ground-hugging grass, and it keeps that humble shape for up to seven years before entering a "rocket stage" and growing four feet straight up, beyond a ground fire's lethal reach.

Something's gone right at Nokuse—bears have come back, ambling in from Eglin Air Force Base and then sticking around. Davis was planning to bring back red-cockaded woodpeckers and, one starry night out on his porch, he spoke to Wilson and me about finding a place for bison. The area's last known woodland bison was shot just before the American Revolution.

"Oh, now you've got me dreaming," Wilson said about the bison. "You've set my imagination on fire!"

Davis's proudest accomplishment was an intense statewide recruitment for a seemingly uncharismatic creature, the foot-long gopher tortoise. Nokuse Plantation director Matt Aresco, who has a PhD in biological sciences, had by then retrieved 3,500 otherwise doomed gopher tortoises from all over Florida. These

Gopher tortoise

tortoises are "ecosystem engineers," a term invented by three ecologists in 1994, meaning they have a transforming influence on their surroundings. It's similar to what beaver families do, but unseen.

Only two-thirds of a longleaf forest ecosystem is visible (trees and ground cover), with the rest underground. At least 360 animal species take shelter in the burrows up to fifty-two feet long and twenty-three feet deep excavated by shy and dusty gopher tortoises. The tortoises retreat down these tunnels to where fires and hurricanes can't penetrate, and where temperatures never sink below sixty degrees in winter or get above seventy degrees in summer. The Florida mouse digs side tunnels, and a tiny, tiny ant lives on the eggs of a spider found only in these burrows.

Gopher tortoises have suffered badly at the hands of both rich and poor. During the Depression, they were dug up and eaten, known back then as Hoover chickens. Nowadays they're buried and simply left there. The sandy soils they dig through are the same soils that developers build on, and gopher tortoises can't dig up, only down, so to kill a gopher tortoise you only have to stop up the tunnel entrance.

In the luminous glow of intense orange Florida sunsets, Davis and Wilson sat on the porch, planning. They pored over maps of nearby industrial timberland that could link Nokuse to the protected half million acres due east, thereby summoning a long landscape, more than sixty miles of continuous longleaf in a grand biodiversity corridor. Davis pointed out there'd be space for even the widest-ranging species, like red wolves and panthers.

Davis died in 2015; his three-hundred-year mission continues. His forest has been permanently protected—and it's still growing. Matt Aresco, now leading the project, has added another 3,600 acres to Nokuse and plants about half a million new longleaf pine seedlings every year. What will happen to the forest landscape east of Nokuse may not be determined for decades—in 2018, Hurricane Michael, a Category 5 storm, flattened, stripped, and denuded 3 million acres of trees across northern Florida and into Georgia. A four-hundred-thousand-acre Panhandle forest has been bought up by the Mormon Church, and much of it is getting converted into a cattle ranch.

It's easier to see Nokuse's future. Aresco thinks that by 2030 some of the trees Davis and he have nurtured will be mature enough for red-cockaded woodpeckers to return. This small bird, whose population has been reduced by 99 percent, makes its nests by excavating cavities in large living pines, preferably longleaf; it does this because the pine resin that oozes around the edges of these hollows wards off predatory tree-climbing snakes. Aresco has rescued an additional 1,500 gopher tortoises.

Aresco is more dubious about seeing bison in Florida, but he's repairing another ecosystem hole by reintroducing, tagging, and monitoring alligator snapping turtles. He calls them the gopher tortoise's "iconic aquatic counterpart." Alligator snapping turtles, along with alligators themselves, are top-of-the-food-chain predators of southeastern rivers and the only living relative of an extinct car-sized turtle. They can weigh 250 pounds, live for a century, and have jaws that close like scissors and a bite that can snap a broomstick in half. Their spiked shells look something like an alligator, but their claws are more bearlike, and they're seldom seen, lying on river bottoms for up to forty-five minutes at a time and wiggling pink tongues that look like worms to lure nearby fish.

Alligator snapping turtles have no known predators except for people. Unlike gopher tortoises, almost eaten to extinction during the Depression, alligator snapping turtle numbers were relatively stable until the more affluent 1960s and 1970s, when they were trapped in massive numbers to be turned into Campbell's canned turtle soup. On one river, three to four tons of turtle meat were extracted every day. Today, Campbell's no longer sells turtle soup.

The wolves howl

GALLATIN VALLEY, MONTANA

Riding around in a pickup truck at the Flying D Ranch in Gallatin Valley, Montana, one July afternoon several years back felt like a return to an unrecoverable past, to the "seens of visionary inchantment" that Meriwether Lewis came across in 1805 when he and William Clark made their way through what would become Montana. Lewis recorded encountering—no spellcheck in the expedition's equipment—"immence herds of Buffaloe, Elk, deer & Antelopes feeding in one common and boundless pasture." We were on the lookout for wolves. I'd settle for bison.

The 113,613-acre Flying D, up at the northwest corner of Greater Yellowstone, is a Ted Turner operation, and only a fraction of the 2 million acres he owns in the United States and Argentina. The ranch has about two thousand elk and five thousand bison. Before the 1870s, it's been said, it would've been easier to count all the leaves in a forest than to count the bison. After fifteen years of mass slaughter, there were only 325 bison left in the nation.

The Flying D is a large-scale, long-term experiment in ecosystem restoration. The premise, according to State Senator Mike Phillips—the pickup truck's driver—is that in ranch country, a wildlife refuge can pay for itself if it's also run as a business. The big bison herd, which replaced a cattle operation, is mostly raised for sale—bison burgers are available at all the Ted's Montana Grill restaurants around the country. Several bull elk on the ranch are hunted annually by high-end outfitters.

Other species are welcomed, celebrated: mule deer, grizzlies, cougars, moose, pronghorn antelope, cutthroat trout, and the occasional wolverine—nearly all the animals that were present before settlers arrived in Lewis and Clark's wake. Wolves found their way to the D in 2002, seven years after being reintroduced to Yellowstone. The D's wolf pack, called the Bear Trap Pack, is the largest in Greater Yellowstone—or was until around 2010, when it got so big it split into two separate groups.

Phillips, a wildlife ecologist, is part of the Democratic minority in the Montana Senate. Since 1997 he's also served as the director of the Turner Endangered Species Fund (TESF). "It's the largest and most significant such family-funded initiative that we know of in the world," he says. I ask him what the D will look like in a hundred years. "Exactly like now," he says with a laugh, "providing we get a good June rain."

Ted Turner was at the D that afternoon for a private meeting with the Greater Yellowstone Coalition and made a point of introducing himself. In jeans and a crisp sport shirt, he seemed quite chipper. "Here's a piece of land," he said, pointing to the high, snowcapped peaks behind him, "that could've been a resort. Twenty-eight minutes from the airport, or downtown Bozeman, or a good Division II football game. But it's perfectly placed as a beachhead for wildness. Seemed to me the choice was obvious, and it's a good thing we stepped in when we did."

He said the Flying D is the largest private property in Greater Yellowstone, a critically important part of this connected landscape. "It's clear nowadays that to protect imperiled species we need to operate at enormous scales that make sense to nature but transcend anything people have assembled," he said. "And it's just as clear that no country will ever have the money to buy up all the unprotected pieces. But it doesn't all have to happen on public land, since private ranches like this one can promote ecological integrity. Private lands are working landscapes; they're money-making businesses. And I think we've invented something entirely new here—call them 'wild working landscapes'—where we make a profit and so does the planet."

Large carnivores, Phillips says, are an excellent lens for looking at landscapes. Their movements and migrations expose broad corridors that already exist. The unanswered question is whether we can develop "socially accepted corridors," as Phillips calls them, along these same routes, so that people within this now-inhabited habitat can coexist with the big creatures in their midst. This is similar to the questions Diane Boyd raises concerning grizzlies. "The GYC folks talk about moving from tolerance to acceptance to appreciation," Phillips says, "though I usually substitute 'admiration.'" Which sounds like Wilson's biophilia in bite-sized, time-released doses.

The bison aren't thundering as we drive slowly through the great herd. They merely stand around massively, impressively, the calves frisky, the bulls larger than our truck. It seems to take forever to get past them. Then we strike off cross-country and uphill. "Let's go howl at the wolves," Phillips says.

A bald eagle perches on a fencepost, and a couple of four-foot-tall reddish sandhill cranes stalk sedately through rolling, grassy slopes filled with purple lupine, white yarrow, and yellow blanketflowers. We stop at a high, sweet-smelling meadow and, once Phillips cuts the engine, an enormous silence envelops us, broken only by the buzzy trill of a song sparrow.

We have to whisper since sounds carry so well in this natural amphitheater. Valpa J. Asher, the TESF wolf biologist with us, says some wolves might show up a mile away. "You'll think they're floating," she tells us quietly. "Wolves are all leg." They would be at eye level halfway up a steep, rocky slope over on the far side of a deep valley. No guarantees, of course. We're looking out at a wolf rendezvous site, a kind of aboveground den, where wolf pups, when old enough, get brought to learn the landscape.

Farther away, the skyline is dominated by the pointed crests of the Spanish Peaks, snowcapped even in summer. There's a rumble of thunder, and it suddenly starts to pour. Wind whistles in our ears. It gets colder, and we retreat to the truck, where Phillips breaks out deli sandwiches and cans of Jamaican lemonade. Then the sun comes out again, and there's a double rainbow to our right. "The D is showing off—this is too cool," Phillips says matter-of-factly, far more restrained than YouTube's Double Rainbow Guy.

Then—there they are. Dots to the naked eye, but vividly close through a spotting scope. A black adult, a gray adult with a black ruff, and six pups, four black and two gray, gamboling, sniffing the ground, chasing one another, dispersing, then regrouping. Definitely floating. Phillips grins, throws his head back, and howls across the valley. On the other side, the two adult wolves throw back their heads and howl. The sounds are faint but unmistakable.

A second-chance landscape

NEW ENGLAND WOODLANDS

New England would seem to be a Half Earth slam dunk, a landscape on the upswing of a yo-yoing transformation. The region was 90 percent forested when the Pilgrims arrived, but almost two hundred years later farmers chopped down all but 20 percent of the trees during a "sheep fever" that can in part be blamed on Napoleon and the stirrings of globalization.

After Napoleon overran Spain in 1808, a Vermonter carried off a herd of merino sheep, prized for their soft, premium-priced wool, which until then had been a monopoly of the Spanish aristocracy. The thirty-year wool craze that followed has been called a mania as powerful as any religious fanaticism. New England's famous stone walls, rocks piled up by hand like the Egyptian pyramids but with more stones than the pyramids, are a remnant of that period. Then this vast series of sheep pens was abruptly abandoned as farmers and herders moved west.

The forests returned, though no one in the twenty-first century will see anything like those first forests' practically sequoia-sized eastern white pines, trees that awed early settlers. Timbering is common in the newer woods, and even if left strictly alone, white pines need a couple of centuries to tower over everything in sight. Even so, the reforests, if they can be called that, instill their own wonder. Self-seeded, they spread again to cover 79 percent of New England.

A 2010 report, "Wildlands and Woodlands," refers to the entire six-state region as a continental-scale habitat corridor. If the pace of land conservation can be doubled, said this study, then by 2060 New England can stay 70 percent forested forever. An updated report calls for a tripling of the pace of conservation because, both reports agree, the area is something rare in the biosphere: a second-chance landscape.

Some environmentalists give this outcome no better than a fifty-fifty chance. Most of the land in New England is in private hands with, in general, larger tracts up north and much smaller holdings as you move south (hundred-, sixty-, or twenty-acre lots). Which means property maps of New England display a fragmented landscape rather than a unified one. No one is proposing turning New England into a national park. What can be done, conservationists say, is to ensure biodiversity on private property by paying landowners to protect present and future forests; in technical terms this is known as a "conservation easement."

Approaching thousands of individual landowners about this, one at a time, could defend and define natural corridors so they remain seamless for animals and plants, setting up formal connections between parcels that previously were, in a legal sense, merely adjacent.

Money is an obstacle—though easements cost less than outright land purchases—and another is finding people to do the paperwork, which traditionally has been handled by small local land trusts. Now these land trusts are amalgamating themselves into larger associations called Regional Conservation Partnerships (RCPs), so as to take on bigger projects.

Driving for several days through the New England woods, I ran across one north-south wildlife corridor, about two hundred miles long. It couldn't be called forgotten because it was never defined, although Thoreau wrote lovingly about one mountaintop up north, Mount Monadnock, in New Hampshire. On a satellite-generated nighttime map of New England, now that such things exist, this corridor pops out unmistakably. These maps show city lights as bright white smears separated by a fascinating absence and emptiness. That dark is where the wild things are.

This column of dark land in the middle of southern New England has a band of light on one side, created by New York and the cities along the Connecticut River Valley, and a splash of white on the other, radiated by Boston and Providence. The dark land itself is a cascade of rolling wooded hills that course down from the White Mountains through New Hampshire, Massachusetts, and Connecticut on their way to the marshes along Long Island Sound.

The corridor, known only as the eastern uplands, has hills that are humbler than the Taconics and Berkshires to the west. It has never attracted a school of painters or their cachet. But because of its intactness, this stretch of the forest—White Mountains to Whitecaps, it could be called, or W2W—is the single decisive interruption in what's now a four-hundred-mile-long line of cities from Washington to Boston, the so-called Northeast Megaregion.

W2W derives much of its strength from an act of brute force. In the 1930s, Boston drowned four towns, evicted 2,500 people, and moved more than 7,600 graves to create the Quabbin Reservoir, a huge, U-shaped lake in the center of Massachusetts. Further development was banned on 56,000 acres of woodland around the reservoir to keep the water pure. Moose, black bears, and bald eagles, all long gone, returned. Anchored by what people have been calling an "accidental wilderness," several Regional Conservation Partnerships focus on land that spreads out from the reservoir. The biggest is Q2C, the Quabbin-to-Cardigan Partnership, whose goal is to protect up to half of the 2 million acres between the reservoir and a peak at the southern tip of the White Mountains.

Outstandingly and even improbably, W2W offers an older, slower sense of countryside that's no longer common in the East, a seemingly endless landscape where towns are like way stations or solitary boats bobbing on what an eighteenth-century geographer called an "ocean of woods." That is what you can see today when looking down from a small plane—a few towns, a few farms, and the ceaseless woods. Chris Wells, who was a Q2C coordinator, grew up in suburban New Jersey, studied planning in Manhattan, and lived in tiny Wilmot, New Hampshire. "There are hawks in my yard," Wells told me. "Bobcats on the front lawn. Some nights I hear coyotes howl—I could be living in the African veldt."

At the Norcross Wildlife Sanctuary, I met up with Dan Donahue, who was director of land protection and stewardship. The sanctuary straddles the Massachusetts-Connecticut border not far south of Quabbin. The core of Norcross's eight-thousand-plus acres has a remoteness to it, a hushed, back-of-beyond quality. The land was bought in the 1930s by Arthur D. Norcross, founder of the Norcross Greeting Card Company, still remembered for popularizing Valentine's Day cards.

Norcross's great interest was rescue work, relocating plants about to be destroyed—including, as he noted proudly, an entire colony of Hartford fern taken from a doomed Quabbin town "before the bulldozer and the flame throwers did their work and the area was flooded." Donahue said, "Mr. Norcross saw this place as an ark. The truth is you can't make an ark big enough to save species. But you can have arcs instead—arcs of land, like the one we're standing in the middle of."

In 2008 there were twenty-three Regional Conservation Partnerships in New England, and now there are forty-four, covering two-thirds of the landscape. A number of conservationists say this never could've happened if Bill Labich hadn't been around to introduce people to one another and get them talking. But Labich, lively and easygoing, a forester and planner and the father of non-identical twins, denies being an RCP whisperer: "I really just get people to collaborate a bit," he told me. Though he will acknowledge his life changed forever in seventh grade, when he saw a filmstrip about a Finnish family working in the woods, pruning a white pine forest. The children gathered the branches and then they all had a picnic. Ever since, Labich has thought about what people can accomplish cooperatively.

These days Labich seems to be always on the job; his title is senior conservationist at the Highstead foundation, in Redding, Connecticut. On a quiet day for him, because he had time to walk me through a 160-acre Massachusetts forest looking for a wood thrush, he told me, "My biggest role is that I can see where we need to be. I stand for the possibility of making that happen." It's as though Labich can beckon others into a future where conservation goals have already been met. In 2015, the RCP coordinators decided to see what could happen if they all worked together for five or ten years on the same project, acting as a network of networks. Could they accomplish something big, perhaps sustain a threatened species?

They chose the wood thrush, a medium-sized cinnamon-brown forest bird with a plump, speckled belly whose numbers have declined 60 percent over the past fifty years. This has been blamed on habitat loss in Latin America, where the birds winter, but a recent study by the Smithsonian Migratory Bird Center points to the need to keep eastern North American forests, their breeding grounds, from being cut over for development. Wood thrushes do best in larger, older forests with densely growing tall trees and thick shrubs that make it harder for crows and jays to see their nests from above. These nests are artfully disguised, often with a trailing strand of grapevine to make them look haphazard and random. The interiors of bigger, deeper woods are also less likely to be reached by ground predators like raccoons and cats that prowl forest edges.

This forest Labich took me to, on the outskirts of Amherst, Massachusetts, met all the criteria. It was part of a continuous woods that stretched west from the Quabbin wilderness only three miles away; in the nineteenth century it was quarried for stones to build Amherst College, but there's been no logging for a hundred years. Labich and I didn't see a wood thrush—they tend to stay hidden—but we found a wood thrush nest. So we were in the right place.

The wood thrush is giving RCPs an opportunity to enlist a new set of partners, the Cornell Lab of Ornithology and statewide Audubon societies throughout New England, eastern New York, and Pennsylvania. It's something of an indicator species—

Wood thrush

meaning that its habitat also suits the needs of other forest birds, including scarlet tanagers and black-throated blue warblers.

Then there's the soft, uncanny, flute-like *ee-oh-lay* song of the wood thrush, one of the first birds heard in the morning and one of the last to go silent at dusk—sounds that stop people in their tracks and once heard are not forgotten. A one-bird duet, it's been called, since a wood thrush can move each of its two "vocal cords" (membranes that act like vocal cords) independently. "Whenever a man hears it," Thoreau wrote, "it is a new world and a free country, and the gates of heaven are not shut against him."

A kind of water heaven

SIERRA VALLEY AND LAKE TAHOE, CALIFORNIA
I was canoeing silently through the heart of the 2,500-acre Sierra Valley Preserve in the northern Sierra Nevada, gliding past

white-faced ibis, large, iridescent wading birds. A group of these birds is called a congregation, a wedge, or a stand. There were also yellow-headed blackbirds, which have golden heads and throats and white on their wings and nest in colonies. A flock of these is known as a cloud, a cluster, or a merl, an old Scottish word.

From the preserve, I looked out across thirty thousand acres of cattails, reeds, and winding waterways, and beyond that to ninety thousand acres of grasses and sagebrush, all of it flat and open and deeply still. Before Sierra Valley became an enormous meadow with a river running through it, this nearly mile-high valley ringed by eight-thousand-foot peaks was a glacial lake the size of Lake Tahoe, the bright blue Sierra jewel fifty miles southeast. In the distance, and I could easily see for miles, were some cows, horses, and old barns, and overhead was a graceful flight of greater sandhill cranes. With no sounds to block them, their honking, rattling bugle calls can be heard two and a half miles away.

The preserve is public open space within the headwaters of the Middle Fork of the Feather River, which supplies drinking water to 20 million Northern Californians. It's a nesting or stopover site for more than 280 different kinds of birds, the greatest concentration and diversity of birdlife anywhere in the Sierra Nevada, in the middle of the Sierra's biggest agricultural landscape. A place of unique juxtapositions where, as gets said locally, sagebrush meets bulrush. Also where the Sierra and two other big landscapes come together, the Cascades of the Northwest and the Great Basin, a place where rain and snow are self-contained and never reach the ocean. Black bears from the Sierra Nevada, and pronghorn from the Great Basin, and even elk and gray wolves from the Rockies—all intermingle.

Still clearly visible in the Sierra Valley is the back end and aftermath of the 1849 California Gold Rush. Before that, the Washoe people, meaning the "people from here," summered in the valley for perhaps nine thousand years; Spanish explorers didn't get this far. The Washoe people suffered greatly as more than sixteen thousand miners swept through the valley. Treaties were ignored, settlements disrupted. Meanwhile, the idea of quick success took on the name "California dream." James P. Beckwourth, the first of the new settlers, was part African American, part white, and part Native American. Later, farmers and their families from Swiss and Italian mountain valleys followed and immediately felt at home. The farms and ranches they laid out, initially to feed the miners, remain in place today; many are now in the hands of their descendants.

The northern Sierra, sometimes called the "secret Sierra" and the "gentler Sierra," is also the vulnerable Sierra, more populated and less protected. It's part of the same range—the mighty Sierra that John Muir described so rapturously: "Miles in height, and so gloriously colored and so radiant, it seemed not clothed with light, but wholly composed of it, like the wall of some celestial city." But in the north the mountains are lower, easier to get to and across, so this is where settlers settled and where the transcontinental railroad came through. As an incentive, the railroad was awarded every other square mile of land for twenty miles, creating a checkerboard landscape of alternating public and private land. This pattern persists.

"As it lay there with the shadows of the mountains brilliantly photographed upon its still surface," Mark Twain said of Lake Tahoe, "I thought it must surely be the fairest picture the whole earth affords." A noble sheet of blue water, he called it. John Muir said Tahoe was "a kind of water heaven," a summing up of all the mountain lakes he'd seen in his lifetime. "Three months of camp life on Lake Tahoe," wrote Twain in *Roughing It*, "would restore an Egyptian mummy to his pristine vigor, and give him an appetite like an alligator. I do not mean the oldest and driest mummies, of course, but the fresher ones."

Congress, after creating three national parks in the southern Sierra, debated making Lake Tahoe a national park on four different occasions. Each time Congress decided that the lake, which is seventy-two miles around, had too much private land to qualify. After World War II, as ski resorts and casinos were added, Tahoe became a center for growth. There was talk of circling it with multilane freeways and building a city the size of San Francisco along its shore.

In the late 1960s, Ronald Reagan, then governor of California, unexpectedly became a Tahoe hero, signing an agreement with his friend and fellow conservative Paul Laxalt, governor of Nevada, to protect the lake's purity and clarity. (Because of a nineteenth-century mapping error, a third of the lake is in Nevada, not California.) "The environmentalists were giving us all these reports," Laxalt later told the *Los Angeles Times*, "that on our watch, any day, Tahoe could turn gray. That got our attention. Rather grimly, we came to the conclusion that a super government agency was the only solution—which was ironical because that went against the grain of everything we stood for."

Growth in the area, though now capped around Tahoe itself, continues. A map based on a 2011 book, *Megapolitan America: A New Vision for Understanding America's Metropolitan Geography,* by Robert E. Lang and Arthur C. Nelson, shows the emerging Sierra Pacific Megapolitan Area. By 2040, they say, Tahoe and the land around it and even Reno, forty miles northeast, is likely to become the northern arm of a cluster of cities and suburbs that reach back to San Francisco, two hundred miles away. The idea is that as the U.S. population expands, many if not most of the newcomers will be absorbed within ten or twelve urban mega-regions. NorCal is another name for Sierra Pacific, though both may be placeholders; they sound like merged corporations, not real places.

Think of the opportunities this foreknowledge could reveal.

What could be saved in the Sierra if you knew NorCal was on its way? Or Southland, for that matter (sometimes called SoCal), a megaregion that by midcentury may run east from Los Angeles to Las Vegas? Or Cascadia or Great Lakes, names that have been proposed for a Seattle-to-Portland megaregion and for one from Chicago to Toronto?

"Every iconic landscape," the leaders of the Northern Sierra Partnership declared in 2018 in the *Northern Sierra Conservation Atlas*, "has its 'now or never' moment, when people who care either step-up to take action, or live to regret their failure to do so." The partnership, a unique, ready-to-be-copied conservation campaign, is a collaboration of The Nature Conservancy, the Trust for Public Land, a group of northern Sierra businesspeople, and two local land trusts. It was set up in 2007 as a "booster rocket" to raise $340 million of federal money and state money, foundation grants, and private donations to protect 188,000 acres of forests, lakes, valleys, meadows, and working ranches, many within twenty miles of Tahoe.

Lucy Blake, president of the Northern Sierra Partnership, told the *Palo Alto Weekly* that it's not easy being a "wrangler," coordinating five separate groups with no sustained history of working together. But already they're halfway to the goal. By the time of my canoe ride, a third of the private land in the Sierra Valley was protected with conservation easements, so it will never be built on. The first valley rancher to place an easement on his property was a man in his nineties. He stood up from his wheelchair at a meeting of landowners, and said this was the only way he could rule from beyond the grave and guarantee what his place would always look like. In the midst of the ranches, the Sierra Valley Preserve, all 2,500 acres, will protect a future for birds and the public's ability to see them up close.

Lake Tahoe has a permanent population of 55,000 and gets

Lucy Blake

15 million visitors a year, about three times as many as the Grand Canyon. The iconic measurement of the lake, taken annually, is the Secchi depth—a way of assessing the water's clarity. It was invented by a nineteenth-century Jesuit astronomer and consists of lowering a twelve-inch white disk, sometimes called a white dinner plate, into the water and seeing how far down it remains visible. Historically there was at least one hundred feet of visibility, but in recent decades it was losing about a foot a year. This was before intervention by the Tahoe Regional Planning Agency (set up by Reagan and Laxalt), which by controlling stormwater runoff and preventing erosion, avoided the loss of another eighteen feet.

The agency has broad regulatory powers but prefers to rely on persuasion rather than coercion. Its aim is to be as careful of the Tahoe basin as if it were a NorCal national park. For twenty years there have been no billboards. "You have to go slow with people to go fast," Joanne S. Marchetta, the agency's executive director told me. "Slow means smooth. And smooth turns into fast." Instead of coming to Tahoe for their own individual slice of heaven, visitors

Joanne S. Marchetta

and residents can help protect the whole of John Muir's "water heaven." Marchetta added, "Ultimately, we're trying to get people to grow a new muscle for adaptation, and for change."

Connecting the dots

GREENWICH VILLAGE, NEW YORK CITY

I took a slow summer walk through my own neighborhood with Timon McPhearson, an ecologist and director of the Urban Systems Lab at the New School. McPhearson got me to see how everything outside that isn't remotely parklike could be the perfect setting and prelude to a park-in-the-making—asphalt streets and cement sidewalks baking in the heat, rows of brownstones, ornate but battered cast-iron fronts and grimy brick walls of old loft buildings, blank plate-glass exteriors of new apartment towers, the dark slivers of space between buildings that almost but don't quite touch.

Behind my apartment building, practically hidden from the

street, is a backyard garden, the remaining open land on what was once the Randall farm. More than two hundred years ago, in a will drawn up by Alexander Hamilton, Captain Robert Richard Randall set up a retirement home here for old sailors, Sailors' Snug Harbor. As the area developed, new buildings generated enough income to create an elegant campus for the "worn-out" sailors over on Staten Island. But the garden, which runs down most of the block, has been carefully restored by the current landlord, New York University. At one end is a small grove of crabapple trees and hornbeams that attracts blue jays and an occasional migrating woodpecker. To this day the garden is a tiny piece of city land that has never been excavated, built on, or paved over.

In front of the building, on ground long since paved over, McPhearson drew my attention not to a concrete planter with bright summer flowers but to a crack in the pavement less than an inch wide between the curb and the sidewalk. It was entirely filled with miniature green plants no more than two inches tall. "Ruderal vegetation," McPhearson explained.

The word "ruderal" is from the Latin for "rubble"; the study of it is actually part of Cold War history. Even before the Berlin Wall went up in 1961, East German travel restrictions cut West Berliners off from access to the surrounding countryside. No longer able to make field trips, local botanists, led by Prof. Dr. Herbert Sukopp, later West Berlin's commissioner for nature protection, turned to inventorying and mapping the "unexpected neighbors": rubble plants springing up in the *Brache*, vast stretches of wasteland in a city 80 percent destroyed by World War II bombing. Sukopp grew up in Berlin and was only fifteen when the war ended. In an interview with Bettina Stoetzer in *Cultural Anthropology*, Sukopp said, "How could I have studied anything other than rubble spaces?"

It was a break with ecology's traditional wilderness focus, a first close look at urban ecosystems. In Berlin it turned out there was more biodiversity in the rubble than in the fields and suburbs just outside the city. Shunned as eyesores, these devastated, desolate landscapes—ruderal areas—could now be seen for what they were, a beneficial part of a healthy ecosystem that lowered summer temperatures and absorbed floodwaters from storms.

These "unruly, tenacious, and opportunistic" plants, writes Sarah Cowles, a landscape designer, in an essay called "Ruderal Aesthetics," "trigger optimism and imagination" about what can happen next. Don't call them weeds, suggests David Seiter, a Brooklyn landscape architect. Call them SUPs ("spontaneous urban plants," also the name of a book by Seiter).

First, you have to see them. McPhearson says that rudimentary ruderal strips, like the one on my corner, tend to go unnoticed. "Plant blindness" is a term coined by a couple of American botanists, James Wandersee and Elisabeth Schussler, in 1998. What causes it? Plants don't have qualities we look for instinctively, either to approach or avoid: they're stationary, often similarly colored, and don't have eyes or even faces. Also, humans tend to prefer to look at things at eye level, or just below.

McPhearson and his students have been working to flood the senses by saturating cities with plants wherever you look—up, down, and sideways. They call this Connect the Dots, a dot being anything already green, from a large park to a backyard garden to a ruderal toehold, the smallest green growing out of a crack in the sidewalk.

It's a 3D approach. Spread green along the ground, and incorporate green roofs, meaning planted rooftops. (Twelve percent of roofs in Germany are green, and thanks to a 2019 New York City law, they're required on all new buildings.) Also, tie ground and roof together with green walls—creating façades covered with cascades of irrigated plants.

Green wall

Some trace this green-wall technology back 2,600 years to the Hanging Gardens of Babylon, one of the Seven Wonders of the Ancient World. As a modern device it dates to 1938 and is credited to Stanley Hart White, a professor of landscape architecture at the University of Illinois, and a brother of E. B. White, the writer. "I guess everyone has crazy brothers and sisters," E. B. White wrote to their mother. "I know I have." White continued:

> Stan, by the way, has taken out a patent on an invention of his called "Botanical Bricks," which are simply plant units capable of being built up to any height, for quick landscape effects, the vertical surfaces covered with flowering vines, or the like. He thinks that the idea has great possibilities for such things as world fairs, city yards, indoor gardens, and many other projects. I think perhaps he has got hold of something.

McPhearson himself grew up immersed in nature, spending a lot of time in a cottage next to a fishing pond in the woods on his family's 160-acre farm in east-central Indiana. He was particularly fascinated by the great horned owls nesting year after year in the hollow of an old oak tree. As an ecologist studying cities, he knows that more than half of the 7.8 billion people on earth live in cities today, and by 2050 more than two-thirds of the 9.7 billion people (the population then) will be city dwellers. His dream for 2050 is that city people will sense their own closeness to wildness.

Several days after McPhearson and I took our walk, a beaver, a species relentlessly hunted for its fur by early New Yorkers, was seen in Manhattan for the first time in more than two hundred years. Some people said the beaver might have been lost or confused. Others say (and I tend to agree) it's a good omen.

Next Steps

While the 2020 Australian bushfires that killed more than a billion animals were still burning, Danielle Celermajer, a University of Sydney sociologist and director of a multispecies justice project, wrote, "As we watch humans, animals, trees, insects, fungi, ecosystems, forests, rivers (and on and on) being killed, we find ourselves without a word to name what is happening. True, in recent years, environmentalists coined the term *ecocide*, the killing of ecosystems—but this is something more. This is the killing of everything. *Omnicide*."

A 2019 letter to *The Guardian* signed by, among others, Greta Thunberg and a former archbishop of Canterbury, began, "The world faces two existential crises, developing with terrifying speed: climate breakdown and ecological breakdown. Neither is being addressed with the urgency needed to prevent our life-support systems from spiraling into collapse." And in a 2020 *Washington Post* article titled "The 2010s Were a Lost Decade for Climate. We Can't Afford a Repeat, Scientists Warn," Kim Cobb, director of the Georgia Institute of Technology's Global Change

Program, sums up the challenge ahead: "Our decisions over the next 10 years will affect the magnitude of climate change for centuries to come. I don't think it can get more sobering than that."

Even if we haven't yet come up with household words for what's at stake, we're already better equipped to defend the biosphere, more so than we might think. A kind of "upskilling" has been going on—a word economists use to talk about getting people ready for jobs that will exist in the future. In terms of forestalling the extinctions emergency, it's a matter of making connections on the ground and in the mind.

The "human predatory pattern" is the name Jessica Thompson, a Yale anthropologist, gives to the fact that humans are the only primate that eats other animals their own size or bigger, a practice she thinks began with our ancestors more than 2 million years ago. I've seen something else at work. Despite humanity's long, deeply entrenched history of killing, we're able to construct a new ability, a human protection pattern that can safeguard the biosphere and its species, something emerging as efforts grand and small converge. This means saving forests like the Boreal, tracing the travels of hundreds of thousands of animals, putting green walls on city buildings, and setting aside half the planet for other species.

Innerview

The upskill of concern for protecting all of life, a way of looking around that combines wonder, possibility, vulnerability, and urgency, is akin to the Overview Effect reported by astronauts looking back at the earth. Benjamin Grant's Daily Overview Instagram account (more than a million followers) posts a high-resolution satellite image of someplace on earth humanity has changed but that few people have seen from above: "We aim to inspire the Overview Effect," he says.

Only this is the Innerview, seen from down here within the biosphere, the sense that the finite infinitude of the biosphere is our shared home and we can act on its behalf and in the interest of its inhabitants. It's why there are already people working on a wealth of different projects; why the United Nations has declared the 2020s the Decade on Ecosystem Restoration; why the 196 countries in the Convention on Biological Diversity will set a goal of safeguarding 30 percent of the continents and oceans by 2030. This is a next step the whole world can take.

By early 2020, 15.1 percent of the planet's land was protected and 7.9 percent of the oceans—both figures far, far short of 50 percent. Wherever you are and whatever your other skills, there are efforts you could support and become part of to help pick up the pace of moving things forward.

These are some of the projects and ideas under discussion or under way in North America and beyond:

Several groups are keeping track of and working on overarching goals. There's the Half-Earth Project, inspired by E. O. Wilson, and the Nature Needs Half movement that Harvey Locke helped launch, whose accelerated goal is "50% by 2030," accord-

half the earth for the rest of life

ing to their website. The Wyss Foundation, in partnership with the National Geographic Society, sponsors the Campaign for Nature—Hansjörg Wyss, a Swiss-born billionaire entrepreneur who lives in Wyoming, said in 2018 that his aim is to protect 30 percent of the world by 2030 ("30x30"), and he's donated $1 billion to bring this about. "For the sake of all living things," Wyss wrote in *The New York Times*, "let's see to it that far more of our planet is protected by the people, for the people and for all time."

The Udall-Bennet Thirty by Thirty Resolution to Save Nature, proposing a national goal of conserving 30 percent of the United States by 2030, was introduced in Congress in 2019 by two western Democratic senators, Tom Udall of New Mexico and Michael Bennet of Colorado. The bill is pending, and the Biden administration has embraced the idea. Canada and more than thirty other countries have made the same 30 by '30 pledge for their own nations as part of the High Ambition Coalition for Nature and People.

As we head toward midcentury, it's becoming clear what a warmer world will look like and feel like. "What will the twenty-twenties bring?" Elizabeth Kolbert asked in a 2020 article in *The New Yorker*. "In geophysical terms," she said, "this question is almost too easy to answer": along with higher temperatures than in any previous decade, droughts will be more "punishing," and there will be more flooding and "ever-higher sea levels." But what about a warmer world where more and more of it is protected? My sense is that even in the midst of profound changes, people will come to see that landscapes dedicated to preserving the rest of life are zones of steadfastness, the sinews of the biosphere.

The opposite ends of the earth

"People need to see big patterns," Paula Mabee, director of the National Ecological Observatory Network (NEON), told *Science* magazine in 2020. NEON is a thirty-year program to answer continental-scale questions about how U.S. "ecodomains" will change by 2049 by integrating data from eighty-one sites from Alaska to Puerto Rico.

Perhaps the biggest pattern of all, at least in the Western Hemisphere, is the one slowly and quietly taking shape just beyond NEON's focus. In 1987, Mikhail Gorbachev, then leader of the Soviet Union, proposed that the entire Arctic become a "zone of peace" and called for international cooperation to develop "a united, comprehensive plan for protecting the environment of the North." The Arctic has been warming twice as fast as the rest of the planet; sea ice is melting so rapidly the area may see ice-free summers by 2040, something that last happened about 2.6 million years ago.

Three decades later, in 2018, Canada, Russia, the United States, and six other countries signed a central Arctic Ocean agreement to ban commercial fishing, for at least sixteen years, from more than a million square miles of the Arctic Ocean, an area larger than the Mediterranean Sea. Parallel to this are negotiations to set up a Pan-Arctic Network of Marine Protected Areas. Still other ideas put forward would create a great park in the Arctic, north of the eighty-eighth parallel, or the Marine Arctic Peace Sanctuary (MAPS), extending over 5 million square miles of the Arctic Ocean.

On land, Northeast Greenland National Park (Kalaallit Nunaanni nuna eqqissisimatitaq in Greenlandic), the planet's biggest national park, more than one hundred times the size of Yellowstone (100 Ys, as it were), dates back to 1974 and covers more than 375,000 square miles, almost half of Greenland. Although Kalaallit has no permanent human residents, there are up to fifteen thousand musk oxen (40 percent of the world's population).

Then there's Antarctica, never even glimpsed by people until 1820. The 1959 Antarctic Treaty set aside—"forever"—that entire continent, all 5.5 million square miles, as an area for scientific research free from military activity or nuclear explosions, a place that "shall not become the scene or object of international discord."

Since 1998, Antarctica has also been declared something more than a demilitarized zone—a "natural reserve, devoted to peace and science." This followed a 1991 agreement to ban mining and oil exploration for at least fifty years, making Antarctica the equivalent, *The New York Times* said, of a "world park." It's the driest and windiest continent, has 90 percent of the earth's ice, no amphibians or reptiles, and constitutes the largest wilderness on earth.

Maybe four thousand people work there in summer and the

number shrinks to about a thousand in winter. In February 2020, the temperature reached 69.35 degrees Fahrenheit at one research station, the highest ever recorded in Antarctica. Despite this warming, the coast and surrounding waters and islands are still home to seals, whales, and more than eight thousand invertebrate species, and every spring are breeding grounds for penguins and more than 100 million other birds. Proposals are pending to create enormous marine sanctuaries in these offshore waters.

One threat to Antarctica's profusion of wildlife is the growing number of human visitors, more than fifty thousand a year. According to Barry Choi, a Canadian travel expert, people are seeking out sights that are under threat. He calls it "last-chance tourism."

In 2016, Kristine Tompkins, a former CEO of Patagonia clothing company, gave a talk at Yale about why she and her late husband, Douglas Tompkins, cofounder of both The North Face and Esprit, spent a quarter century buying land for national parks in Chile:

> When you're dealing in large landscapes, the number-one thing you have to do, before you leave or kick the bucket, is get it so that the citizenry itself has fallen in love with and therefore become protective of their national park system. That takes maybe a generation, a generation and a half. A park's a huge money-maker, but much more important, it becomes a point of pride. And then if some knucklehead comes along, which they do every so often, and attempts to fill the edges of, say, Olympic National Park, people will go berserk.

Two years later, Tompkins Conservation, the nonprofit set up by Doug and Kris, gave a million acres to the Chilean govern-

ment. Chile then added 9 million acres of state-owned land to create five new national parks and enlarge three others. More than simply adding stand-alone parks, this act gave the country a connected system of parks, an Emerald Necklace, to use Olmsted's phrase for the nineteenth-century Boston parks he helped create. Only in this case the system, the Ruta de los Parques de la Patagonia, is 1,700 miles long and links seventeen parks with 28 million acres spread out along scenic roads and waterways that reach down through Patagonia to the southernmost tip of South America, Tierra del Fuego, the *fin del mundo*, the end of the world.

Many of the parks are wild places with ice fields, glaciers, fjords, and waterfalls. Some had been ranches overgrazed by cattle and sheep and are now once again home to herds of guanacos (wild relatives of llamas) and the pumas (mountain lions) that hunt them. "There is something about the expanse of Patagonia, a kind of haunting soulfulness to it that affects you physically," Tompkins told a *New York Times* reporter in 2018. Once local people develop a sense of ownership and pride, they can become the first line of defense for conservation and "throw the brake on the extinction crisis," a Tompkins colleague said to *The Guardian*.

Permanent protection at and near the poles can become a kind of twenty-first-century force field that can emanate throughout the hemisphere. From an Innerview perspective, if both ends of the globe become safe havens for life, then the great remaining job is to fill in the middle.

Crossover conservation

North America, thanks to its lucky geography, has a built-in advantage when it comes to thinking bigger about conservation— good bones, with two great mountain ranges, the Rockies in the West and the Appalachians in the East. It's where many of the wild things still are and where most of the people aren't. But everywhere—in open countryside, in farmland, in urban megaregions, and in other cities and towns—the tasks are the same: Protect what survives; repair what's been damaged; connect places that have been severed; collaborate as never before; and make the whole effort personal by bringing the natural world back into people's daily lives, wherever they are. All this in a continent that will be home to over 100 million more people by 2050, with less space for them. According to a University of Southern California study, sea-level rise could displace 13 million coastal U.S. residents by 2100.

The key to making all this happen will be "crossover conservation," a coordinated effort to create broader coalitions, with most of the work (by far) getting done by people who don't have the kind of money available to Hansjörg Wyss or the Tompkinses. There's the 2,000-mile-long Yellowstone to Yukon Conservation

Initiative, the Network for Landscape Conservation, and the Center for Large Landscape Conservation. The Wildlands Network, cofounded by Michael Soulé, Reed Noss, and others, connects people who connect wildlands.

These groups promote a new understanding, citing the need for biologists and activists to work with one another over long distances, to save species by finding corridors that can link sundered landscapes and habitats with Indigenous communities, ranchers, farmers, hunters, homeowners, and local governments. David Johns, another Wildlands Network cofounder, calls this "creating little orchestras" around a common vision. A proposed Wildlife Corridors Conservation Act would find corridors on federal land and support "comprehensive corridor network projects" on state, tribal, and private land.

An even bigger partnership would integrate the biosphere into actions bringing climate change under control—these are known as "nature-based solutions." There's a series of studies in *Science Advances, Nature,* and *Nature Ecology & Evolution;* one of the contributors is James E. M. Watson, an Australian biologist who is mapping wilderness areas around the world. The studies show that protecting wilderness can cut the world's extinction risk in half, and that intact forests, like the Canadian Boreal (the Fort Knox of carbon), slow global warming by absorbing a quarter of all the carbon added to the atmosphere. First, Watson says, save the forests that are still extensive, unbroken, and undiminished. For Watson these are earth's "crown jewels."

In 2015, British ecologist Tom Crowther and a team of researchers in Zurich calculated that the world has 3 trillion trees and room for 1.2 trillion more, a further foresting that could reduce carbon in the air by another 25 percent. The Trillion Tree Campaign, an extension of what had been the Billion Tree Campaign, even caught the attention of President Trump in 2020, though

Crowther cautioned it would take a century for these new trees to gain the absorptive abilities intact forests already have.

Crowther has since turned his attention to the carbon in soils, telling *New Scientist* that "they certainly contain more than the planet's vegetation and atmosphere combined" and that there's enormous potential to encourage "the world's soil to accumulate more. Not just in forests, but beneath grasslands, in peatlands and even on farms." Decarbonizing the air by turning away from fossil fuels, as championed by groups like 350.org, cofounded by Bill McKibben, can be complemented by recarbonizing the land, saving the landscape that's whole and restoring what's broken, as stimulated by drives like 1t.org (the abbreviation standing for "1 trillion trees").

All-species design

Ronald L. Mace was an American architect who contracted polio at age nine and had to be carried up and down stairs to his college classes, and whose lifetime of work led to the Americans with Disabilities Act in 1990, prohibiting discrimination against people with physical and mental difficulties. Mace coined the phrase "universal design." This refers to buildings accessible to everyone, and to tools and equipment anyone can use without difficulty. Examples include shower stalls without doors, water fountains at different heights, and kneeling buses with lowered ramps so people in wheelchairs can roll aboard.

Another outsized partnership for a Half Earth world would unite people inspired by the natural world with those who construct the built world, to find ways to open up the concept of universal design and bring it closer to the even more inclusive goal of "all-species design."

Up to 230,000 birds are killed in New York City every year, flying at full speed into buildings. They think they're seeing sky when in fact they're looking at windows or glass walls that reflect the sky. In 2019, by a vote of forty-three to three, the City Council passed a bill requiring nonreflective "bird-friendly" glass on all new buildings—tinted or imprinted with dots. The Jacob K. Javits Convention Center, whose mirrorlike façade had made it one of the city's deadliest, installed this kind of "fritted" glass in 2013, and bird deaths, according to New York City Audubon, went down 90 percent. San Francisco and Oakland have similar laws; in other cities, plans are in the works.

It's light, not windows, that—along with habitat loss and pesticides—is deadly to insects. Of the millions of insect species, many still unidentified, perhaps half are active at night, and light pollution affects a quarter of the planet. Entomologists refer to this as ALAN (artificial light at night), and a 2020 paper in *Biological*

Bird-friendly glass at the Javits Center, New York City

Conservation, by entomologists from the United States, Canada, Australia, and New Zealand, reported that a third of insects circling a lightbulb they may have mistaken for moonlight will die before morning, either exhausted or eaten.

The "fatal attraction" of car headlights instantly kills 100 billion German insects every summer, part of a rapid global insect decline that has sometimes been called an "insect apocalypse." Fortunately, as Brett Seymoure, from Colorado State University, a coauthor of the ALAN study, told *The Guardian,* light pollution, whether it comes from a streetlight or a gas flare, is relatively easy to handle. Streetlights can be aimed so they only shine down and not up into the night, and superfluous lights can be eliminated. "Once you turn off a light," Seymoure said, "it is gone. You don't have to go and clean up, like you do with most pollutants."

Providing animals with highway overpasses and underpasses that restore free movement is probably the most advanced built-world/natural-world partnership. What has yet to be added is a parallel focus on highway drivers, so they get a sense they're not just between places but always within some part of the biosphere. Interstate 84, for instance, cuts through what seem like endless woods in northeast Connecticut. This is part of the still-intact and green White Mountains to Whitecaps landscape and passes within half a mile of Snow Hill in Nipmuck State Forest. Every June, Snow Hill is covered with pink and white and sometimes almost-purple blooms, the small, seemingly quilted cups on mountain laurel bushes that can grow twenty feet high. Signs on the interstate would help bring this to mind; so would audio guides that could be picked up by car radios. All part of a more general push, a kind of bio-scaping, to make highways more like parkways.

The Trump administration's wall between the United States and Mexico is the great exception to this de-barriering of the

Mountain laurel in full bloom

landscape. An eighteen- to thirty-foot-tall steel-slat wall is only one component of what U.S. Customs and Border Protection has been calling a "new border-wall system." In 2020, Christian Alvarez, a CBP spokesperson, told National Public Radio, "So it's not just gonna be the barrier itself."

Alvarez talked about a 150-foot-wide graded and graveled— meaning clear-cut, no vegetation—"enforcement zone," the width of the right-of-way for many interstate highways, with powerful floodlights, surveillance cameras, and "an all-weather access road." Now estimated to cost $18 billion up front, long-term upkeep and repairs on the project, if completed, "could put taxpayers on the hook for billions of dollars," according to *The Washington Post*.

A *BioScience* paper signed by more than 2,500 scientists (including E. O. Wilson) says the wall system will threaten the

future of 1,506 plants and animals, 62 of which are already critically endangered. It would cut through seven conservation areas
in Texas, including Big Bend National Park and the Lower Rio
Grande Valley National Wildlife Refuge, and would split the
National Butterfly Center into two pieces, leaving almost seventy
acres on the Mexican side.

The National Wildlife Federation called the border-wall system "one of the biggest potential ecological disasters of our time,"
and NWF president Collin O'Mara suggested the border could
instead be secured by a "virtual wall" made up of underground
pressure sensors, seismic detectors, radar, and drones that would
have little impact on wildlife. "We have more effective and less
harmful solutions today," he said, "than what the Chinese came
up with thousands of years ago."

Mr.—or Ms.—President, tear down this wall.

Hand in hand

In 1966, when he was in his late eighties, Benton MacKaye came
up with his last large landscape vision, a "spiel and scheme,"
he called it, "for combining nationwide Wilderness Trails with
nationwide Wilderness Areas." As MacKaye's biographer, Larry
Anderson, has pointed out, MacKaye sent the secretary of the
interior a four-page memo and a hand-drawn map showing what
are now the Pacific Crest National Scenic Trail (2,653 miles long
down the length of the Cascades and the Sierra Nevada) and the
Continental Divide National Scenic Trail (3,100 miles following
the crest of the Rockies). These "wilderness ways" would have as
their "stopping points" or "way stations" more than a dozen wildlands between Canada and Mexico.

It was time—"a logical next step" MacKaye's memo said—to

make his original idea for the Appalachian realm into "a series of Nationwide wilderness ways." Even this was only a beginning, he felt, since the Continental Divide Trail, or Cordilleran Trail as he called it, could also stretch into Canada, as "a first step toward 'dealing with continent-wide problems.'" MacKaye wrote to a friend, "The Trails can help the Areas, and the Areas can help the Trails." He wrote to another, "These two efforts . . . should work hand in hand."

Actually, MacKaye may have provided the inspiration for the Pacific Crest Trail—in 1926, Joseph T. Hazard, a hiker who made a living selling textbooks, met Catherine Montgomery, a supervisor at a teachers' college in Bellingham, Washington, on one of his sales calls. This story is told in Hazard's 1946 book, *Pacific Crest Trails*. Montgomery was herself a hiker, and one day, out of the blue, she asked Hazard, "Why do not you Mountaineers do something big for Western America?" Taken aback, Hazard wanted to know, "Just what have you in mind, Miss Montgomery?" Her answer: "A high winding trail down the heights of our western mountains with mile markers and shelter huts." And she told him about an article she'd seen describing plans for the Appalachian Trail.

Peter Parsons, a resourceful outdoorsman, was a Swedish immigrant who'd jumped ship in Oregon in 1909, when he was twenty, to escape a vengeful captain. Parsons had never heard of Benton MacKaye. But in 1924, two years before Montgomery and Hazard met, Parsons shouldered a rucksack labeled HEADING NORTH—MEXICO TO CANADA, and hiked the entire length of what is now the Continental Divide Trail.

With or without MacKaye's influence, much of his "spiel and scheme" has come into being—and much of it still has not. There's a National Trails System with eleven National Scenic Trails and nineteen National Historic Trails, one of them the Lewis and

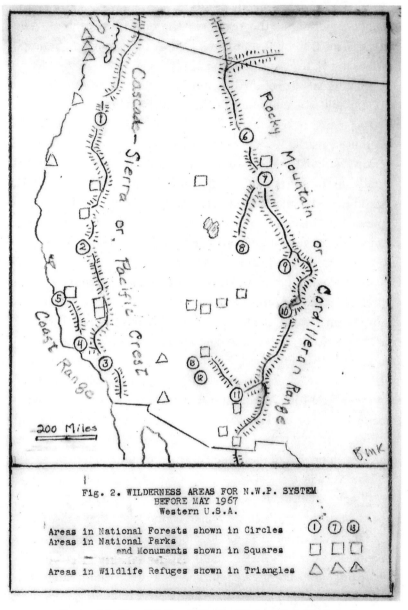

Fig. 2. WILDERNESS AREAS FOR N.W.P. SYSTEM
BEFORE MAY 1967
Western U.S.A.

Areas in National Forests shown in Circles
Areas in National Parks
and Monuments shown in Squares
Areas in Wildlife Refuges shown in Triangles

Benton MacKaye's "scheme" for the western United States

Clark Trail, approximately 4,900 miles long, which retraces the route they took as they explored the northwest from 1803 to 1806. The Trail of Tears follows the path taken by one hundred thousand Native Americans forcibly relocated west of the Mississippi after passage of the Indian Removal Act in 1830.

As the Partnership for the National Trails System reported in 2019, there are and there aren't 55,000 miles of national trails because "nearly 20,000 miles—or more than one third of the entire National Trails System—have never been built or are not accessible for the public." In 2017, after twenty-five years of effort, hundreds of Canadian communities, working together, completed the Great Trail. Coast to coast, it's 14,000 miles long, and one of the largest volunteer projects in Canada's history. But only about 4,900 Great Trail miles are off-road.

As trail gaps get filled in, can these projects take on an even more MacKayean and Half Earth purpose? So that trails, as they're lengthened, can benefit the landscapes they pass through? R. Travis Belote, a hiker and landscape ecologist with The Wilderness Society in Bozeman, Montana, thinks answers are at hand. In a 2020 *BioScience* article, Belote, along with several coauthors, proposed creating, as an intermediate target, a system of "well-connected protected areas" by 2030 that would cover 40 percent of huge areas like the Intermountain West between the Rockies and the long line of mountains formed by the Cascades and the Sierra Nevada.

"Looking ahead," Belote said when I spoke with him, "trails can be the anchor by which we create connectivity that will persist. For instance, the land along ridgetops that animals need and that people love can be seen as equally necessary to protect, instead of as two separate stories we tell. There's a rule of thumb that conservation biologist Paul Beier, who's been designing migration corridors for a quarter century, came up with in 2018. He said the question he's most often asked by wildlife advocates is, What's the narrowest corridor width that won't be regretted later on? Two kilometers, he says—a mile and a quarter. That way there's plenty of room for animal pathways and for human trails, like the Appalachian Trail, which has a thousand-foot-wide corridor. With a mile and a quarter, animals get the protection they need, and the hiking trails pull in political backing."

It's easy to find organizations that exist to strengthen trails in the United States, a country with more than 47 million people who consider themselves hikers. Virtually all the long-distance national trails have their own support groups, as do many of the shorter National Recreation Trails—almost 1,300 of them, up to 485 miles long, though some are less than a mile. More than 24,000 miles of old rail lines have been converted to hiking and biking trails; the half-finished Great American Rail-Trail will stretch from Washington, D.C., to Washington State. The American Discovery Trail, also cross-country, this one more than 6,800 miles long, runs from Delaware to Point Reyes National Seashore, north of San Francisco, passing through cities, farms, and wildlands.

In the East, the Appalachian Trail is no longer the only mountain path reinforcing the integrity of MacKaye's Appalachian realm. In 1952, Earl Shaffer, the first thru-hiker on the AT, suggested stringing together trails on the western slopes of the Appalachians.

Since 2007 it's been the mission of two dozen hiking clubs, calling themselves the Great Eastern Trail Association, to braid these trails into an 1,800-mile route paralleling the AT, leading from Alabama to upper New York State. Already the AT has an urban counterpart to the east, the East Coast Greenway; three thousand miles long, it connects 450 communities from Maine to Florida and gets 15 million visitors a year. Some hikers say this opens the way for myriad crosscutting east-west "ladder trails" to link all three north-south trails (the Great Eastern, the AT, and the East Coast Greenway) into a single eastern mosaic.

Three R's

City residents are far closer than they might think to wildlands that could use their help. In the one hundred largest cities in the United States, 84 percent of the parkland is woodlands, wetlands, and other natural areas: 1.7 million acres all told, nearly the size of Yellowstone. Even New York City, despite four hundred years of almost nonstop growth, has 7,300 acres of forests, and these forests have their own champion, the Natural Areas Conservancy, which considers them as important to the city as the museums, theaters, and libraries. The conservancy has launched a first ever twenty-five-year plan to keep these forests wild and well.

Some urban groups think beyond city limits—for instance, Chicago Wilderness, whose *Biodiversity Recovery Plan* covers nearby parts of Wisconsin and Indiana as well as the Chicago suburbs. Advocacy by Openlands, also in Chicago, made it possible to transform an old army arsenal southwest of the city into the twenty-thousand-acre Midewin National Tallgrass Prairie. In Texas, Houston Wilderness pledges to protect and preserve 24 percent of the land throughout an eight-county area by 2040. In Portland, Oregon, the Urban Greenspaces Institute works in coordination with more than one hundred other groups across a 2,850-square-mile region that reaches into neighboring Washington State. The UGI motto: "In livable cities is preservation of the Wild."

Even more ambitiously, the Green New Deal would have a profound effect on the United States. As Nicholas Pevzner, a

landscape architect, points out in *Landscape Architecture Magazine*, shifting to an economy no longer based on oil, gas, and coal means "connecting the sunny and windy parts of the country" to cities along new electric lines marching across "thousands of miles of American landscape, likely sparking strong local opposition, as power line projects tend to do."

Unless, Pevzner says, this is seen as an opportunity on a "grand territorial scale" by making sure, with a MacKaye-like sweep of the mind, that these new corridors are also designed as trail networks to "connect Americans to their nearby wildlands." This is *American Progress* updated and reimagined in honor of the Green New Deal.

On a more personal level, people can ask themselves, as Ole Amundsen urges, Who is the Henry David Thoreau or the Andrew Wyeth of their area? Or, for that matter, the Lucy Braun or Stephen Kakfwi? Who is thinking bigger? What other species call this place home? How can the Innerview be supported and strengthened? The three R's: what must be retained, what can be restored, and how it could all once again be reconnected.

Spark Bird

The word "endling," only coined in the mid-1990s, means the last of its kind, the final surviving plant or animal of a species. Passenger pigeons, which decades earlier had darkened the North American skies, became extinct after Martha, their endling, died in the Cincinnati Zoo in 1914. Lonesome George, a 101-year-old Galápagos Islands giant tortoise, was an endling, the sole survivor of his Pinta Island subspecies. A worldwide sym-

Lonesome George

bol of conservation, he is memorialized in a Galápagos museum almost desperately named the Hall of Hope. Lonesome George, "Solitario Jorge," died in 2012.

For birders, a spark bird is the bird that reeled them in, the one they heard or saw in a never-to-be-forgotten moment when they were taken unaware by its color or shape or song or flight. Thereafter, they were forever open to the splendor and everywhereness of birds. The spark bird marks a beginning, despite the presence of endlings.

For many people, it doesn't have to be a bird that changes how they see the planet. Or leads them to thinking about the earth at a more comprehensive scale. Or what inspires them to reconfigure the landscape in a way that won't let species vanish.

Here are the spark bird moments for some of the people in and out of these pages:

As a boy, Bill Finch, a botanist, caught sight of a firecracker plant in full brilliant bloom in the Alabama woods.

One afternoon in Florida, M. C. Davis, a businessman, attended a lecture about the needs of black bears, just to escape a traffic jam.

As a young man, Steve Kakfwi, now a Dene elder, lay down on the ground near the Arctic Circle and simply allowed the land to seep into his bones. He had recently left the prison of a Canadian residential school.

"I was very young," says Sarah Charlop-Powers, executive director and cofounder of the Natural Areas Conservancy. "My parents were tenant organizers and we lived in a squat in the South Bronx, during the height of the 'Bronx Is Burning' era. At that time, the New York Botanical Garden was free and open to the public. My mom used to load up our station wagon with as many neighborhood kids as she could fit and drive the few miles to the garden. She would say, 'You can't know what season it is if you can't see the trees.'"

"I was home from college," says Genevieve LaRouche, project leader for the Chesapeake Bay Field Office of the U.S. Fish and Wildlife Service. "This was the end of summer, and I was in a car full of friends hanging out in the Walt Whitman mall parking lot on Long Island. I happened to see a historical marker noting that this was near the famous poet's birthplace. I'd been reading Whitman at the time, enthralled with his joyful celebration of nature and humanity. It dawned on me that Whitman would be horrified to see his name associated with this soulless parking lot, the ugly paved miles, and the unthinking consumerism. I was filled with anger and a conviction to do something—to protect land and be more thoughtful about how we live on the planet."

Steve Kallick, a lawyer and environmentalist, drove north when he was seventeen, and drove north some more, and reached the end of the roads. He then met up with the immensity of the Canadian Boreal Forest and saw a chance to set history right.

When young, Reed Noss, a biologist, found the skull of a passenger pigeon in Ohio, and from then on stood in the shadow of extinctions.

"As a kid in the Colorado Rockies, I saw two coyotes walking across a meadow," says Jodi Hilty, president and chief scientist of Y2Y. "One coyote was limping quite badly. The other coyote went ahead, caught a rodent and brought it back, and offered it to its injured companion. The moment was just so stunning and so special, to 'see' the otherwise private life and relationship of these two wild creatures. I was enraptured, and have been ever since."

Greta Thunberg, the young Swedish environmental activist, told David Attenborough on the BBC that watching his nature documentaries "made me open my eyes for what was happening with the environment and the climate," and that "made me decide to do something about it."

In grade school Bill Labich, a New England forester and planner, saw the power of cooperation when he watched a filmstrip about a Finnish family pruning a white pine forest.

Under a scorching sun, Anna Valer Clark, an artist, gazed out at parched Arizona land and felt compelled to turn it green.

"One of my earliest memories," says Emily Bateson, director of the Network for Landscape Conservation, "is being out on a lake in the fabled Adirondack Mountains in summer, the lovely quiet of it all. The oars dipping in and out of the water, leaving little whirlpools that mesmerized me. On occasion, an oar would catch on a lily pad, pulling up the long, wet stem and delicate white flowers, always tantalizingly out of reach. As we skimmed along the water, the beavers would slap their tails, the loons would greet us with their beautiful haunting trill, and it all just felt like home."

"I was four years old," remembers Joanne Marchetta, executive director of the Tahoe Regional Planning Agency. "My parents decided to show me and my siblings the country via a trailer trip across the United States. There was one moment I recall vividly. After many days of flat prairies and plains, we approached a long, straight ribbon of roadway extending to the horizon in front of the car, and my small self had to scrunch down in the seat to see the full expanse in front of me, the Teton Range in Wyoming rising miles high. I was awestruck. Somehow I knew I was seeing something I did not understand, the sublime and unknowable power of the earth and its nature."

Benton MacKaye, a forester and philosopher, swayed from a tall tree on a Vermont mountaintop and caught a view of the distance, and the future.

My own moment happened more slowly, cumulatively, if you can say that about a moment. Over the course of this book, I came to realize that for this glorious, beleaguered planet, there's room enough and time enough.

Acknowledgments

I've been buoyed through the process of getting this book together by a group of exceptionally talented people at Knopf—my editor, Ann Close; her assistant, Todd Portnowitz; the production editor, Ellen Feldman; the text designer, Maggie Hinders; the jacket designer, Chip Kidd; Nicholas Latimer, director of publicity; publicist Jessica Purcell; and marketer Morgan Fenton. They brought in the equally formidable copy editor, Amy Ryan. David Lindroth created the stunning large landscapes maps, and Dana Tomlin, the father of GIS, digitized the potential protected landscapes map. Many thanks as well to Reagan Arthur, Knopf's publisher, and to LuAnn Walther, editorial director at Vintage. What a superlative team!

Robert Sullivan, a compellingly good writer, looked at early drafts, and Ben Kalin, a fact-checker to dream of, double- and triple-checked the text, always with aplomb.

And I'm always grateful to be working with the best agent in the business, Amanda Urban of ICM.

I've been lucky enough to get to know some of the extraordinary people working to protect half of North America and stave off the mass extinction crisis, and many of them are mentioned in the book, notably E. O. Wilson, who was also kind enough to write the introduction. Many others were equally helpful and need to be cited here, such as Kathleen Horton, E. O. Wilson's assistant, and Paula J. Ehrlich, president and CEO of the E. O. Wilson Biodiversity Foundation. Vance Martin is president and Amy Lewis is vice president of the WILD Foundation, the group that sponsors Nature Needs Half, and Ruth Midgley works with Harvey Locke on post-2020 conservation targets.

At the outset I got guidance from old friends who've thought about conservation planning for decades, among them Robert D. Yaro, president emeritus of the Regional Plan Association, and Peter R. Stein of The Lyme Timber Company. I also benefitted greatly from talks with Ethan Carr, professor of landscape architecture at the University of Massachusetts Amherst; Kyle Copas, communications manager at the Global Biodiversity Information Facility; Eric Dinerstein, director of biodiversity and wildlife solutions at RESOLVE; Joel Dunn, president of the Chesapeake Conservancy; David R. Foster, president of Highstead; John Griffin, a former secretary of natural resources in Maryland; Walter Jetz, professor of ecology and evolutionary biology at Yale; Jane Lawton, chief development and communications officer at Forum for the Future; Dan Murphy from the Chesapeake Bay field office of the U.S. Fish and Wildlife Service; Stuart Pimm, professor of conservation ecology at Duke; John Polisar, jaguar program coordinator for the Wildlife Conservation Society; Dennis Shaffer, director of landscape conservation at the Appalachian Trail Conservancy; Mike Slattery, Chesapeake coordinator for the U.S. Fish and Wildlife Service; Michael Scott from Resources Legacy Fund; and Don A. Weeden, executive director of the Weeden Foundation.

Carol Kavanagh from the Indigenous Leadership Initiative helped arrange my trip to the Boreal Forest. David S. Wilcove, professor of ecology and evolutionary biology and public affairs at Princeton, introduced me to Martin Wikelski. Brian B. King, publisher at the ATC, helped me on many matters, and Carol Paterno welcomed me to the Appalachian Trail in Pawling, New York.

In New England, Peter McKinley, research ecologist and conservation planner at The Wilderness Society, toured me through the High Peaks of western Maine, and I got great help from Kristin DeBoer, executive director of the Kestrel Land Trust; Steve Golden, who spent thirty-eight years with the National Park Service; Spencer R. Meyer, senior conservationist at Highstead; Charles Tracy, superintendent of the New England National Scenic Trail; and Leigh Youngblood, executive director of the Mount Grace Land Conservation Trust.

Elsewhere around the country I was sustained by the thoughtfulness and generosity of Hunter Armstrong, deputy director, and Tessa O'Connell, communications manager, at New York's Natural Areas Conservancy; Caroline Byrd, former executive director of the Greater Yellowstone Coalition; Howard R. Permut, former president of Metro-North Railroad; Ron Sutherland, chief scientist, and Susan Holmes, U.S. federal policy director, for the Wildlands Network; Charles McKinney, former principal urban designer with the New York City Department of Parks and Recreation; Karen Votava, founding executive director of the East Coast Greenway, and Dennis Markatos-Soriano, the current incumbent; and Kara Hartigan Whelan, vice president of the Westchester Land Trust.

Gary Paul Nabhan, ethnobotanist and agricultural ecologist, and H. Ronald Pulliam, former director of the Eugene P. Odum School of Ecology at the University of Georgia, made my trip to Mexico possible—and they were excellent traveling companions and guides.

For help in thinking about greenline parks I'm particularly indebted to Amy L. Freitag, executive director, and Elizabeth Wolff, program director for the environment, at the J. M. Kaplan Fund; and to Brenda Barrett, editor of *Living Landscape Observer;* Billy Fleming, director of the Ian L. McHarg Center for Urbanism and Ecology at the University of Pennsylvania; and Destry Jarvis, a former assistant director of the National Park Service. And people on the ground in Delmarva who generously went out of their way to show me around included Janet Christensen-Lewis, chair of the Kent Conservation and Preservation

Alliance; Kate Hackett, executive director of Delaware Wild Lands; and Darius L. Johnson, director of communications at the Eastern Shore Land Conservancy.

In 1987 a National Park Service team in Philadelphia, comprised of Joan Chaplick, J. Glenn Eugster, Margaret Judd, Cecily Corcoran Kihn, Elizabeth Lukens, and Suzanne Sutro, produced a national map, "Potential Protected Landscapes, Local and State Landscape Areas," that documented landscape conservation efforts under way in the continental United States. I am especially grateful that this map, a blueprint for greenline parks, appears in this book.

All along the way and at each step I've been lifted by the transforming editing of my wife, Lois.

Index

ILLUSTRATION CREDITS

Tony Hiss is the author of fifteen books, including the award-winning *The Experience of Place*. He was a staff writer at *The New Yorker* for more than thirty years, a visiting scholar at New York University for twenty-five years, and has lectured around the world. He lives with his wife, the writer Lois Metzger, in New York City.

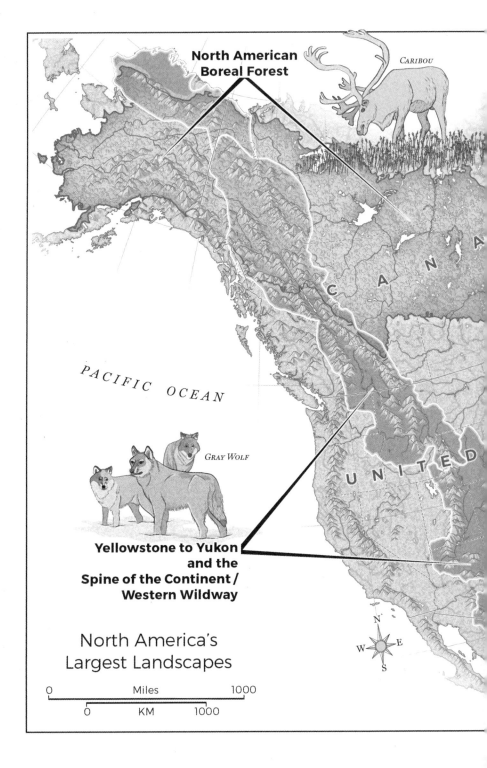

North American Boreal Forest

CARIBOU

PACIFIC OCEAN

GRAY WOLF

C A N A

U N I T E D

Yellowstone to Yukon and the Spine of the Continent / Western Wildway

North America's Largest Landscapes

N
W · E
S

0	Miles	1000
0	KM	1000